Automating Humanity

Automating Humanity

Published in the United States by powerHouse Books,
a division of powerHouse Cultural Entertainment, Inc.
32 Adams Street, Brooklyn, NY 11201-1021
e-mail: info@powerHouseBooks.com
website: www.powerHouseBooks.com

First edition, 2018

Library of Congress Control Number: 2018954636

ISBN 978-1-57687-920-7

Printing and binding by Friesens Corp., Altona, Manitoba, Canada

10 9 8 7 6 5 4 3 2 1

Printed in Canada

 powerHouse Books

To my mom and dad, who have supported me throughout my life, regardless of the fact that they never understood what I was trying to do. Well, now you don't have to guess anymore. Now you know.

And to my grandpa, who my family lost to dementia during the writing of this book. The mind's ability to both uplift and destroy worlds without any visible signs of damage is truly incredible—a truth we should all keep in mind as we move into this modern era of mental economics.

We will miss you, Grandpa, but the things you have taught us about hard work, faith, and goodwill towards others will never be forgotten.

Table of Contents

Part 1: WTF Is Happening?!

Traditional industries based in physical reality are being swallowed whole, wiped from the face of the Earth without any visible replacement. Communities are disappearing into the screens they're glued to. News organizations are sure robots are going to kill us all and there's no way to stop them. The world is on fire and nobody knows what's going on.

Part 2: How Did We Get Here?

We've been struck by a tidal wave of digital technologies that is ripping through our world like a chainsaw cutting through butter. Everything's happening so fast most people don't even know where it started or how it got this way to begin with. What's happening seems unreal. But it is real. And it's here to stay.

Part 3: What Could the Future Be?

We're now capable of doing things previously only possible in sci-fi films. Although there's plenty to be concerned about with the current state of technology, the overwhelming majority of what's happened has been proven to be good for society. And the good news is that if we do this well, there is an abundance of benefits still to come.

Part 4: How Do We Get There?

Despite the threats created by modern technologies, plenty of good has been done as well, which should not be understated or downplayed. However, if we want to make the future enjoyable, we must discover ways to amplify the good parts and curb the bad.

Automating Humanity

Joe Toscano

 powerHouse Books

Brooklyn, New York

Welcome to the Future

Automating Humanity is an insider's perspective on everything Big Tech doesn't want you thinking about—from how addictions have been installed at a global scale, to how profits are driven by fake news and disinformation, to how and why artificial intelligence will take jobs at an unimaginable pace. Welcome to the future. You're late.

Between the speed at which modern technologies are evolving, the mosaic of information available on the web, and the pressure to get things done faster than ever before, predicting the future of the tech industry can be challenging. But as a consultant for Google, where my team and I spent our days overseeing what was happening in the industry, forecasting the future, and crafting strategic plans to keep Google stay on top, this was my job. This was a unique position to be in, enlightening in ways that are hard to explain.

What I learned is that we are entering an era of ambient computing, driven by mass amounts of data flowing through artificially intelligent, connected, digital ecosystems. We've advanced so far, in fact, that we're beginning to replicate human thought, perception, equilibrium, and many of the most fundamental aspects of human life in automated systems that are capable of monitoring the world, consuming data, and acting on the information they receive on our behalf—all of which happens inside an algorithm that operates on a cloud server somewhere in the middle of, well, nobody knows. However, despite the fact that many people around the world are unaware of what's happening, these products are scaling the globe, defining the way people experience the world, and impacting society in ways that are difficult to realize until years later, long after the damage has been done.

In June 2017 I left my job consulting for Google due to ethical concerns with the industry—not just Google. While there are plenty of great things happening in the Valley (many of which do not get the press they deserve), there are also plenty of things that get overcomplexified, which keeps them from being discussed. This book is my attempt to change that—to show you the inside of the industry; critically assess the problems; and define an ethical approach to creating a better future. While you will definitely stumble across some industry jargon, I've worked tirelessly to overcome this problem by introducing metaphors and stories that relate historic examples to what's happening today, and paired them with simple graphics that visualize what would otherwise be invisible, abstract concepts. My hope is that outlining the current state of the industry and its impacts in this way will not only improve our collective understanding, but also change the way work is approached and cause more to be demanded from the companies running the marketplace.

In Part 1: WTF Is Happening?! we'll start off by establishing a baseline understanding of what's been happening in the industry. There will be no secrets given away, but there will more than likely be plenty of information you've never considered before—either because it's new to you or because you haven't thought of it this way. This part will put you in a better position to understand why things are happening the way they are.

In Part 2: How Did We Get Here? we'll walk through a brief history of information, technology, and evolution in order to give you a better understanding of just how fast we're moving relative to human history, and why, which will allow us to have a larger conversation about the critical factors underlying these systems.

In Part 3: What Could Happen we'll take a trip to the future and come to understand the potential these systems have to create positive impact. Yes, there are plenty of things to be afraid of but, if done well, the tools being created are capable of relieving us of economic effort, which would allow us to focus on more meaningful aspects of life. It's important to understand these positive aspects if we're going to have a reasonable discussion about how to clean up the industry.

Finally, in Part 4: Moving Forward we'll wrap up with a conversation about what needs to be done to make the industry safe, in which we'll walk through everything from how to understand what a data monopoly is to how to reintroduce the stability that's been drained from our world thanks to Silicon Valley. This part will attempt to define what should be put into legislation versus what should remain a matter of consumer demand so that we can peacefully navigate this transition period without doing catastrophic damage to the global economy.

No matter how ambitious this may sound, do not mistake this for a philosophical text. If I learned anything while consulting Google, it is that any discussion about creating the future should directly relate to the bottom line. What you are about to read will give you insights that will allow you to create a future that is not only ethical and sustainable but also economically savvy. By the end, I hope you have a better understanding of the role you play in this system and how you can leverage that position to help it flourish.

PART 1

WTF Is Happening?!

Traditional industries based in physical reality are being swallowed whole, wiped from the face of the Earth without any visible replacement. Communities are disappearing into the screens they're glued to. News organizations are sure robots are going to kill us all and there's no way to stop them. The world is on fire and nobody knows what's going on.

1

The Race To Zero

The goal of a capitalist system is to be as efficient as possible—to do more with less. However, logic would tell us that eventually there won't be more to do less with. Have we reached that point? Will it be reached soon? A current analysis of the market demonstrates that we are, in fact, headed that way.

In 1930 John Maynard Keynes, a British economist whose ideas fundamentally altered the way governments understood and practiced macroeconomics, gave a lecture titled, "Economic Possibilities for our Grandchildren," in which he speculated on the future of economics and what the rate of technological progress might mean for future generations. In it he coined the term "technological unemployment"—a state of increasing unemployment due to technological progress that exceeds our ability to discover new jobs. During the lecture, he noted:

> *There is evidence that the revolutionary technical changes, which have so far chiefly affected industry, may soon be attacking agriculture. We may be on the eve of improvements in the efficiency of food production as great as those which have already taken place in mining, manufacture, and transport. In quite a few years—in our own lifetimes I mean—we may be able to perform all the operations of agriculture, mining, and manufacture with a quarter of the human effort to which we have been accustomed.*
>
> *For the moment the very rapidity of these changes is hurting us and bringing difficult problems to solve. Those countries are suffering relatively which are not in the vanguard of progress. We are being afflicted with a new disease of which some readers may not yet have heard the name, but of which they will hear a great deal in the years to come—namely, technological unemployment. This means unemployment due to our discovery of means of economising the use of labour outrunning the pace at which we can find new uses for labor.*
>
> *But this is only a temporary phase of maladjustment. All this means in the long run is that mankind is solving its economic problems. I would predict that the standard of life in progressive countries one hundred years hence will be between four and eight times as high as it is today. There would be nothing surprising in this even in the light of our present knowledge. It would not be foolish to contemplate the possibility of a far greater progress still.*[1]

Ahead of his time, Keynes was not only aware of how technological progress impacted society in his own time but also realized that this trend was not going to slow down. With the onset of the digital age, the impact of technology on society has only become exponentially greater. Things are moving faster than ever before. Never before in the history of business have we seen so many industries disappear. Never before in the history of business have we thought it would be intelligent to discover creative ways to make something free. And never before, in the history of business, have humans been considered so worthless. Looking at the world nearly 100 years after Keynes's statement, it's clear the future he predicted is now.

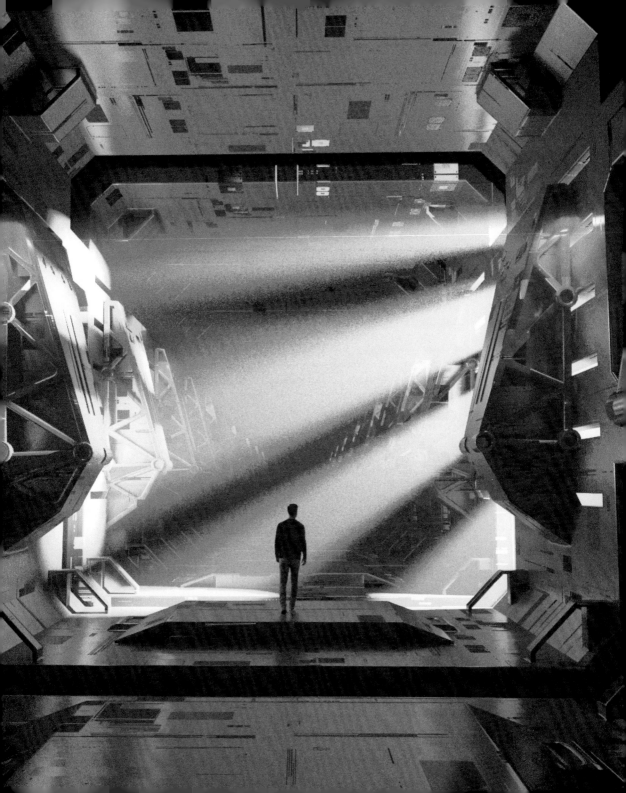

Zero Infrastructure

The cost and risk involved in starting a business has dramatically changed over the past 20 to 30 years, offering new opportunities to many business people. Historically, setting up a business required some storefront, warehouse, or other form of physical infrastructure that forced business owners to settle down and work to pay off the debt they had incurred to acquire the building. Then to grow, a store owner had to construct additional buildings, which increased costs and, more generally, risk—a cost of doing business.

But now, in the digital era, thanks to the internet, the only infrastructure businesses absolutely need is a website, which can be made, essentially, for free in contrast to the traditional cost of establishing a business. The website can then represent a business's storefront, staff, and inventory. It can be accessed 24 hours a day, 7 days a week, 365 days a year, around the globe. And, thanks to translation services, it can be found in almost any language. The only people doing work are the consumers, who must look through options, decide what they want, and then place orders. Sure there are some costs to a business: the cost of purchasing a machine to work on and a domain name to host its site, on top of any other necessary web services, but these costs are nothing compared to purchasing land, building infrastructure, and employing people.

Zero Employment

Today, entrepreneurs leverage computing power, internet access, and code to automate as much as possible. This allows them to do more, with less—much, much less. For example, let's start off with Facebook, a modern, digital media company that, at the time of this writing, is valued at nearly $585 billion; an incredible amount of money, more than the GDP of many nations combined. Yet the company only employs a little over 25,000 people, many of which reside in Menlo Park.[2] Compare these numbers to Disney, a very successful, traditional media company valued at about $159 billion while employing nearly 200,000 people worldwide, and the differences in efficiency become blatantly obvious, resulting in devastating disparities of wealth and power.[3]

But these are media companies. Can the disparity be as great between two companies that make physical goods? Let's compare Tesla and Ford. In just 14 years of existence, Tesla is already valued around $60 billion on the stock market, about $15 billion more than Ford, which has been in business for 114 years to Tesla's 14. This holds true despite the fact that Tesla only produced around 100,000 vehicles in 2017 while Ford produced over 6.5 million. Moreover, Tesla does this while employing a meager 33,000 people, most of whom are located in the Silicon Valley area, compared to Ford, which employs over 200,000 around the world.[4] [5]

Surely there has to be some industry where this is not the case, right? What about retail? Retailers need employees to deal with customers, stock the shelves, deliver goods. The retail industry must be an outlier, right? Here to Amazon and Walmart. When compared to most tech companies Amazon actually employs quite a few people—over 560,000, according to reports.[6] They do this while reaching a market cap near $845 billion. And Walmart? Well, Walmart is valued around $249 billion and employs over 2.3 million people worldwide.[7] This means Amazon has more than tripled the value of Walmart in half the time, with less than a quarter of the employees—a total that is dwindling, thanks to artificially-intelligent robots.

At the end of 2016 Amazon stated it staffed 45,000 android workers in its warehouses. In the first half of 2017 that number nearly doubled to 80,000, then rose to 100,000 by the end of the third quarter.[8] Assuming Amazon maintains speed, we can estimate that by the end of 2017 that number will continue to rise, putting Amazon's robot staff number in the 110,000 to 120,000 range. These robot workers represent the future of unemployment for hundreds of thousands of warehouse workers around the world, Amazon just has a head start. In fact, 2017 was the first year that Amazon's robot staff grew faster than it's human retail staff, which ended up taking a significant cut for the first time in the company's history.

Despite the fact that the number of Amazon's human employees dropped in 2017, Amazon's stock rose nearly 60 percent that year. This occurred because once Amazon replaces humans with robots, it has employees that are able to work 24 hours a day, 7 days a week, 365 days a year without complaining, without healthcare benefits, without paid time off, without asking for raises—without almost all costs necessary to employ human laborers. These are business efficiencies that haven't been seen since the days of slave labor, and investors know the value they bring to a company.

This automated operation also saves Amazon money by improving the company's logistical capabilities. Whereas the average human thinks categorically (i.e., plates, silverware, pots, pans in the kitchen section; shirts, pants, socks, underwear in the clothing section), a robot can think logistically (i.e. based on longitude and latitude). This means if Amazon discovers certain items which don't normally fit in the same category are commonly bought together, they can then store those items close to each other because a robot doesn't care about what types of items they are, it only cares about the location.

So, for example, if Amazon notices that people tend to buy peanut butter when they buy brooms, Amazon can store peanut butter and brooms right next to each other instead of in two separate sections of the warehouse, which means less time is wasted putting together an estimated 3 million orders every day.[9] Even if this only saves Amazon 10 seconds per order on average, at a rate of $13.50/hour this saves Amazon more than $40 million a year.[10] It's undeniable how valuable this is for a corporation focused on dominating competition.

Even the sharing economy will begin to find its perfect efficiency by eliminating the need for human workers. Today, Uber has to pay drivers, but those drivers are not employees of Uber. They are not on salary; do not get benefits; and, in fact, don't even get a car, gas, or service. Uber drivers are independent contractors , and soon—once

Traditional Production vs Tech Production

A comparison of market capitalization values for businesses that operate under traditional production models versus modern technology production models.

■ Tech Production

▪ Traditional Production

Market Cap Value, in billions

Number of Employees

1 T

750 B

500 B

250 B

100,000   200,000   300,000   400,000

Uber gets its self-driving technology operating without bugs, which it has been publicly testing since 2016—they, too, will be eliminated. Consequently, in the future, zero-employment may happen as use of artificial intelligence becomes more common. We can see proof of this if we look at the number of employees in tech companies and the value they generate compared to the number of employees in traditional companies and the value they generate.

We tend to hold a false assumption that bigger companies must mean more employees, but this is a very traditional belief that needs to be overcome. If we look at the data it's clear to see that in modern tech companies there's a point where having more human employees actually becomes a liability—a point in history where human workers are no longer a necessary part of the equation. This means that now, instead of competing against others like themselves, humans are competing for jobs with artificially intelligent machines.

Zero Downtime

Another benefit of the new, digital economy is that companies can reduce the amount of time required for product development cycles. This is because businesses have gone from creating rival goods (things that are not easily replicated and only capable of being used by one person at a time, like a book), to creating non-rival goods (things that can be replicated with the click of a button and used simultaneously by millions of people, like an eBook). So, unlike in traditional businesses where products are shipped to consumers, used for a while, then analyzed to understand customer satisfaction, everything connected to the internet is constantly tracked and analyzed. Businesses can immediately understand how consumers use their products and utilize that information to improve the products in near real-time. This has also dramatically lowered production costs.

Today, modern companies don't care about their product being wrapped up under the Christmas tree, they care about keeping us glued to a data collecting-machine as long as possible. The resulting data then tells them exactly how to improve their products and, in many cases, what products they should create for you before you even realize you need it. This leads to efficiencies in businesses like nothing we've ever seen before. As artificial intelligence becomes more powerful and businesses get more creative, we'll begin to see products that not only fix, but improve themselves, and, not long after, become fully automated organizations that operate without human intervention, outside of initial setup and intermittent maintenance.[11][12]

Artificial Intelligence Is the New Electricity

According to a Gallup report in March 2018, more than 85 percent of people in the United States already use AI in one form or another, 79 percent of whom say it has a positive impact on their life.[13][14] The forefront of this movement to embed intelligence into machines is focused on big, commercial industries, but it will quickly move into every part of life. As we begin to connect non-traditional physical devices to the internet, we're constructing a world that looks and acts like nothing we've experienced before. And although it may feel crazy at times, what we're experiencing today is only the beginning. This transformation has silently taken over our world and isn't looking to slow down any time in the near future.

      Some technologies, like electricity, have become so deeply embedded in our lives that we are not aware of their impact until they are not operating. Soon artificial intelligence will be like electricity. It will be in everything, but go unnoticed until it breaks—like how we only care about our electrical current when a lightbulb goes out. This change won't just be an upgrade in how we interact with and share information, it will fundamentally alter what it means to be human. It will change how we measure our impact on the world; how we learn about, react to, and adapt to new discoveries; and, ultimately, how we understand ourselves. As we move forward, business owners, investors, financial advisors, product innovators—anyone with a stake in the future of a company—can stubbornly hold out in hopes that their industry will not be impacted by artificial intelligence, but the only place that will get them is left behind.

CHAPTER 2

Data Over Everything

Machines experience life through data. It is what allows them to learn and evolve. Without it, artificial intelligence would be highly unintelligent. Because of this, companies are manipulating the fundamentals of life in a pursuit to mine more data.

Everyone who owns a smartphone and regularly uses it is carrying the world's greatest tracking device. It knows where they've been, who their closest friends are, what they look like, what they like to do, what they buy and where they buy it, as well as a long list of other intimate information. And it knows all of this with incredible accuracy because it contains years', if not decades', worth of such information, which can be used to assemble a historical narrative of that individual's life. In fact, even if people turn off their device location in the settings they are still being tracked through cell site location information (CSLI). Generally, this system is only used to coordinate calls and track people who are lost, but law enforcement can get access to it almost anytime they want.

The legality of this tracking is currently being debated in the Supreme Court, where a team from New York University's Data & Society is arguing that cell phones have become a necessity to modern life. Whether used for keeping in touch with family, enabling a higher level of public safety, meeting the demands of modern employment, or a number of other things, the world depends on cell phones to operate at this point—a statement that is especially true in the US where more than 95 percent of adults have cell phones.[12] Their claim is that because owning a cell phone has become so necessary, this continuous tracking of CSLI data violates the public's Fourth Amendment right. But if you think this is bad, it's only going to get worse.

Over the next five to ten years, surveillance-based products are only going to become more popular. Anything that can be connected to the internet and made to be "intelligent" will be. And in order to operate most effectively, these tools also need to understand the context of people's lives, which means they need a lot of data so the system can understand what's going on. Because of this, the greatest minds in technology are now working to discover creative ways to collect information without appearing creepy.

While these technologies will undoubtedly bring many conveniences to peoples' lives, they will also make it easier for companies and governments to track our behavior in ways they couldn't before. Although most consumers don't understand the speed or the scale at which this is happening, the businesses operating in this space do, and they will use it to their advantage. Not only will they monetize the data, they will also use that data to predict what the market wants and drive demand. What they're doing is not so different from what happens in the basement of psychology departments across the country, but the scale at which they're doing it is unprecedented.

In clinical studies of human behavior, scientists are capable of finding statistical significance within a sampling of just a couple of hundred participants. "Big Data," a buzzword that's been floating around in the media over the past few years, allows companies to do the same thing but on a global scale, for corporate profit. However, unlike results from clinical studies where participants know they're being analyzed, data from internet-connected devices represents what social scientists call "behavioral residue"—behavior that has been tracked or acquired unknowingly, reflecting unmodified, unbiased data—like when someone leaves the toilet seat up and everyone knows who used it last.

Also, unlike studies run in college basements that often result in data being collected on college students, the internet has been collecting data from billions of people around the world for the past 20 to 30 years. The amount of knowledge contained in this data is something no one individual or large group of individuals will ever be able to analyze alone. This is a task that's destined to be the job of a machine, and those who are capable of making the machines that can do it will find an unending amount of hidden treasures capable of changing the world. With enough data they can even begin to shape the world as they please. For example Netflix, Spotify, and many other streaming services use data to make content they know people will like. Google, Facebook, Twitter, and other information curation services use your data to show people exactly what they think they want, when they think they want it. Amazon does the same with retail knowledge. This means there is no longer risk involved in their business model. Instead, they can tell people what to like by nudging them towards decisions. Used strategically, this allows companies to shape the market in ways that are most profitable for them. To make this happen, companies need a lot of data though.

Anyone can copy and paste the code from Google and recreate Google. The same goes for any code-based system—Facebook, Netflix, Spotify, it doesn't matter. But without the data these companies possess, their systems would just be nothing like what we experience today since data is what allows machines to understand the world. Just as children learn from their life experiences over time, machines also need experiences to understand how to operate. The difference is that a machine's experience can be instantly plugged in, taking it from novice to expert in almost no time, unlike children who require years or decades of learning to become experts. And the more data points available, the smarter and more "experienced" the machine will be. In this sense, data can be thought of the metaphorical "life experience" for machines. And with this understanding it should be clear why so many people are saying data is the new oil. It is a necessary ingredient to any intelligent system—the fuel to our artificially intelligent machines—and for this reason, companies will do everything possible to monopolize their share.

Mining of the Minds

One thing many people outside the tech world don't understand about this new data era is that cash is no longer the desired form of currency. But those involved understand it well. To them, data is worth more than gold. However, companies don't need money to mine data from people's minds, they have investors for that. What they need is people's attention. And the more attention people pay to the system—the more they click, scroll, and generally remain engaged—the more data can be collected. Consequently, attention has become the new currency. This is a concept that was reinforced by Tim Cook, CEO of Apple, in 2014 when he made a comment directed at companies like Facebook and

Google, stating, "When an online service is free, you're not the customer. You're the product."[3]

The problem with attention-based business models is that there is only so much time an individual can give, making attention a limited resource.[4] The difference with attention, as compared to say coal or oil, is that attention's upper limit has less to do with its physical limitations and more to do with the total number of people multiplied by the number of hours they are capable of being engaged. Because of this, there are two business strategies underlying attention-based companies: get more people online and give them their time back

Getting the World Online

According to Robert Tercek, author of *Vaporized: Solid Strategies for Success in a Dematerialized World*, every year for the rest of the decade cellular companies expect to sell at least a billion smartphones. Some will be replacements for existing, outdated phones, but about half will be first-time users. That means 500 million new internet users each year, for the rest of the decade—a pace that won't slow down until everybody has one.[5] But despite how it may look from the outside, telecommunications companies aren't the only people trying to get consumers glued to a screen.

Google is working on Project Loon, a Wi-Fi distributing network of balloons that travel the edge of space pushing internet down to developing nations. Facebook is doing the same with Aquilla, remote controlled drones bigger than a Boeing 737 that circle the stratosphere in flocks hoping to "bring the world closer together."[6] These companies are not doing this to create informed citizens, stop civil wars, or assist with myriad other problems; they're doing it because they have saturated their native markets and need to onboard more users if they want to generate more revenue. If you think that's cynical, just look where their money comes from; advertising accounts for more than 90 percent of the revenue generated by both companies.[7][8] Being able to reach more people means they can sell ads for more money because they can tell advertisers they'll reach more people.

Amazon, who released an app called "Internet" to India in early 2018 is doing the same in hopes of increasing the number of Amazon customers.[9] This sounds sneaky, but it's not the first time such an offering has been made. Facebook also tried with Free Basics, but the government of India saw Facebook's motivation differently, rejecting Facebook's offer to market what it considered a discriminatory version of the internet, cleverly designed to force people to use Facebook so the company could collect data on its people.[10] In response, Marc Andreessen, an early investor in Facebook and partner at the powerful Silicon Valley investment firm Andreessen Horowitz, sarcastically tweeted: "Anti-colonialism has been economically catastrophic for the Indian people for decades. Why stop now?"[11] He later deleted the tweet, but it had already made Facebook's motivations clear. These companies and the people running them can say things like this and keep their thrones because of the power they possess, but follow the paper trail

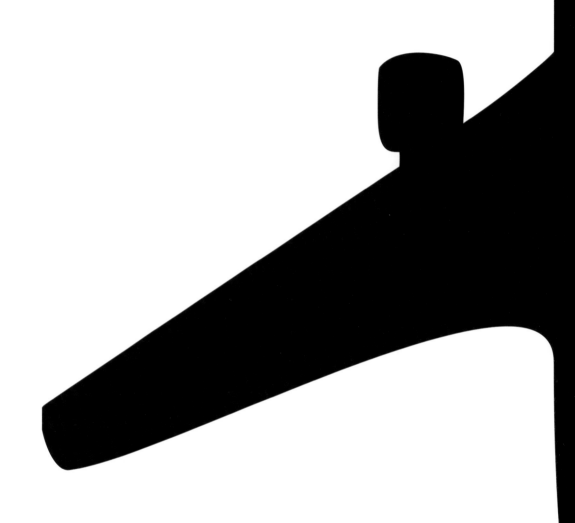

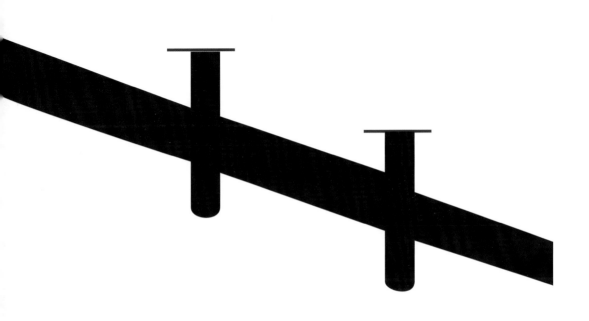

Facebook's Project Aquila has a wingspan that's wider than a Boeing 737, weighs less than 1,000 pounds, and only uses about 5,000 Watts of power—about as much power as three hairdryers—which it acquires from the solar panels on its wings as it cruises the stratosphere for up to 90 days at a time.

back to where the money's coming from and it's easy to understand the true motivation behind their behavior.

Manipulating Time

The second, less obvious way these businesses succeed is by giving people their time back. In order to keep users glued to their screen longer, these businesses work tirelessly to eliminate time consuming activities from users' lives—things like work, family time, shopping, driving, and more. This means any businesses operating in the attention economy are also masters of time manipulation. For example, the reason so many tech giants are interested in creating self-driving cars is less about public safety than one might think, though that's what their PR teams will tell us. Think a little deeper though, and what's clear is that by eliminating the need to drive, people also get their time back. Companies know that if they can eliminate the time people spend paying attention to the road, they will have more time to pay attention to their product.

This may sound cynical, but I've listened to executives in the industry speak about the future of automated cars. "Think of the possibilities that could come with self-driving cars," they say, "Consumers could shop more, they could stream more videos, there would be more opportunity to advertise products. And, more generally, there would be more time for us to engage consumers." Although different leaders may have slightly different opinions or ways of saying it, they're all getting at the same point. At first this may not sound dangerous, but that's because this modern economy is invisible, which makes it difficult to understand. What's being paid in this new economy isn't money but attention, which has a direct connection to our health and freedom. If we're not careful, both will silently disappear from our lives.

Data Is the New Oil

The final aspect of this new economy that we need to consider is how to assess the value of data. For example, what's the value of 500,000 personal addresses? How much should your address cost compared to Oprah's address? How much should it cost if it's packaged with data about your credit card purchases? Or what if it's packaged with your medical history? Nobody knows. There is no general agreement as to its value, unlike resources like oil where we know the price per gallon. Even the people running these companies aren't 100 percent sure how to value their data assets.

The truth is we're in a transition era. We're all still learning and adapting to this new economy. And because of this we lack an agreed upon value for our new currency. But in a world obsessed with data, somebody's going to find a way to profit off the misalignment of society. We can already see this happening in the stock market, where the value of Big Tech companies is inflated beyond reality. But this inflated value is just one of many issues facing us today.

Large amounts of data also cost a lot of money to acquire, manage, and maintain, yet nobody fully knows how to adjust the balance sheet for these intangible costs. Even if accountants are capable of balancing the tangible costs of storage and maintenance, they're incapable of including data assets on the company's books because there's no agreed upon value like there is for other items in the world. This also means insurance companies are incapable of covering lost, stolen, or otherwise misplaced data, which means that if the servers crash, so does the business, and there's nothing anyone else can do about it besides start from scratch.[13]

The only beneficiaries of this misalignment of society are the companies who own the data. As long as there is no agreed upon value for data, these companies can charge whatever they want. This new economy has pushed us into a world created purely of perceived value that even the greatest minds on Wall Street struggle to comprehend, but it isn't stopping companies from moving forward. Modern data empires will do whatever it takes to dig deeper into the data wells of society. The question left to answer is: *How do we value our time and what are we willing to pay?*

CHAPTER 3

The Price of Free(dom)

As the world becomes digital, we're no longer paying for the tools and systems we use with money. Instead, we're paying with attention. This means once companies have gotten people online and given them their time back, it's time to keep them engaged. But is this business model a good thing? What are the impacts of everything being free?

On page 13 of his book, *Hooked: How to Build Habit Forming Products*, Nir Eyal states that his goal is to provide readers with "a deeper understanding of how certain products change what we do and, by extension, who we are."[1] This claim may seem inspiring or disturbing, depending on how it is interpreted. Continuing on, Eyal's writing takes the latter tone as he suggests that well-designed, habit-forming products should create "a feeling that manifests within the mind and causes discomfort until it is satisfied. The habit-forming products we use are simply there to provide some sort of relief."[2] This suggestion corresponds directly to the language used by the American Psychological Association when it refers to addictions:

❝ *Addiction is a chronic disorder with biological, psychological, social and environmental factors influencing its development and maintenance... Heightened desire to re-experience use of the substance or behavior, potentially influenced by psychological (e.g., stress, history of trauma), social (e.g., family or friends' use of a substance), and environmental factors (e.g., accessibility of a substance, low cost) can lead to regular use/exposure, with chronic use/exposure leading to brain changes.*

These brain changes include alterations in cortical (pre-frontal cortex) and sub-cortical (limbic system) regions involving the neuro-circuitry of reward, motivation, memory, impulse control and judgment. This can lead to dramatic increases in cravings for a drug or activity, as well as impairments in the ability to successfully regulate this impulse, despite the knowledge and experience of many consequences related to the addictive behavior.[3]

Consumer device addiction is a well documented phenomena. According to a 2013 study, 79 percent of smartphone owners check their device within 15 minutes of waking up.[4] Another study found that the average smartphone user checks their phone more than 150 times per day, or more than once every ten minutes.[5] Worldwide, 84 percent of people say they couldn't go a single day without their mobile device.[6] In fact, to avoid having their phone taken away, 55 percent of Americans said they would give up eating out for a year, 46 percent say they would work an extra day each week, and more than one-third of the population worldwide would give up sex for a year.[7] People have become so addicted to their devices that in 2017 and early 2018 prominent leaders in the tech community began stepping down in order to speak publicly about what they've done. One of these was former Facebook president Sean Parker who was very honest in an interview with Axios:

❝ *The thought process that went into building these applications, Facebook being the first of them...was all about: How do we consume as much of your time and conscious attention as possible?... And that means that we need to sort of give you a little dopamine hit every once in a while, because someone liked or commented on a photo*

or a post or whatever. And that's going to get you to contribute more content, and that's going to get you...more likes and comments... It's a social-validation feedback loop...exactly the kind of thing that a hacker like myself would come up with, because you're exploiting a vulnerability in human psychology... The inventors, creators—it's me, it's Mark [Zuckerberg], it's Kevin Systrom on Instagram, it's all of these people— understood this consciously. And we did it anyway.[8]

Later in 2017 Facebook's former vice president of user growth, Chamath Palihapitiya, echoed this sentiment to a room of Stanford students, whom he told:

" *Your behaviors—you don't realize it but you are being programmed. It was unintentional, but now you got to decide how much you are willing to give up.*[9]

This is not, however, limited to Facebook or even social media companies, in general. These addictive experiences are implemented at the device level as well. Tony Fadell, former senior vice president at Apple during the creation of the iPod and iPhone, has also admitted to having regrets about the impact of the devices they created:

" *I wake up in cold sweats every so often thinking, what did we bring to the world? Did we really bring a nuclear bomb with information that can—like we see with fake news—blow up people's brains and reprogram them? I know when I take [technology] away from my kids what happens. They literally feel like you're tearing a piece of their person away from them—they get emotional about it, very emotional. They go through withdrawal for two to three days.*[10]

Unfortunately this trend of addictive experiences isn't on track to change anytime soon if we leave it up to market capitalism because the more people use them, the more data companies can collect, which means the more money they can generate. There's absolutely no financial incentive to make their systems less addictive, and that's the real issue here. With these tactics in place and no regulation to enforce protective standards, morals and ethics have gone out the window in pursuit of the dollar bill. As chilling as it sounds, the truth of the matter is that we now live in a world where the demise of humanity has become what's best for the bottom line.

Postural Deformities

Although it often goes overlooked, one of the most visible effects of our devices is the way they are impacting us physically. Whether this is from sitting at desks all day, studying content on tablets, or surfing social media on smartphones, we're now seeing postural deformities in young children and adults that traditionally were only seen in older people with advanced forms of osteoporosis and other debilitating conditions. The effects of modern desk work have gotten so bad that experts are now claiming that sitting is the new smoking, and needing to be hunched over, glued to a screen is a big part of that.[11] After-work screen time is just compounding the damage.

These deformities are caused by the way our heads act like a weight for our neck to hold all day. When we're upright, it's not bad since we've evolved to stand upright and keep our necks straight. But when we bend our neck to look at our phones, tablets, and other devices the story is different. Typically when using our devices we angle our necks about 60 degrees, and with the average human head weighing in at about 10 pounds, this creates a force equal to about 60 pounds on our necks.[12] Over time, our bodies adapt to this poor posture, resulting in lengthened neck muscles and chronic pain. No matter how ready we are to give into our machines, our bodies are not.

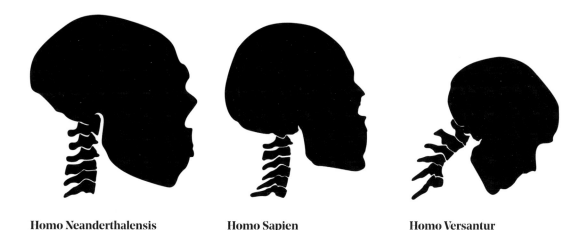

Homo Neanderthalensis   **Homo Sapien**   **Homo Versantur**

In an attempt to heal, our bodies naturally release calcium towards these agitated areas.[13] But if nothing changes, the calcium starts to build up and become permanent. Then, instead of helping, our natural healing mechanisms actually make everything worse. As this build-up turns into chronic neck pain we start to get headaches more often as the pain moves up our spine. Then, not long after it's moved up our spine, the pain begins to head in the opposite direction, down our body, into our shoulders, lower back, and everything else below. These changes in posture then continue on to negatively impact our emotional states as our emotions begin to reflect what our body is already saying.[14] As a species we will continue to pay for our prolonged engagement with machines until the behavior changes us permanently.

Sleep Disorders

Another negative effect of our devices is their disruption of sleep. Until 2015, sleep deprivation wasn't even close to being in the Mayo Clinic's top five health-related issues, but that year it became number two.[15] According to the CDC, more than one third of people in the United States don't get enough sleep. This is especially true of younger populations, who need sleep the most. Whether we are responding to emails until late into the night, binging on Netflix until we pass out, or hunting endlessly for the bottom of our never-ending timelines, the blue light emitting from our devices radically alters the sleeping habits of millions around the globe.

What's happening, according to studies in both the United States and Korea, is that the blue light emitted from our devices significantly suppresses the amount of melatonin released by our bodies, which helps us fall asleep at night. This throws off our circadian rhythm, tricking our body into thinking it's still daytime—not time to go to bed. It has been demonstrated that this effect gets worse when we keep our bedroom lights on than when we use our devices in the dark, but most of the damage is done by the blue light emitted from our devices.[16] This lack of sleep then leads to people dozing off behind the wheel, performing worse at work, and unable to focus during the day. It has also been linked to increases in anxiety, depression, and other mental disorders, which means lack of sleep is only catalyzing the effects of our psychological issues.[17] Scientifically speaking, there are no benefits to continuing this relationship.

Loneliness and Depression

The last negative effect we'll cover here (although there are many more we could talk about) is the damage being done by the social-validation feedback loops embedded in these systems. An overwhelming majority of the research coming out is showing that these systems are doing incredible damage to our mental health. What we're seeing across the world, especially within the younger generations, are significant increases in loneliness and depression complemented by a steep decrease in mental health.

As we begin to define ourselves by our internet relationships and the way we're represented online, the amount of likes, comments, and shares we get chemically affect us in the same way as social interactions in real life. A like sends the same dopamine rush we get when friends compliment our new shoes. A comment sparks a chemical reaction similar to the feeling we get during a personal conversation. And a share makes us feel like our friends are promoting us to the world.[18] However, the same holds true in the opposite direction. In the absence of likes, comments, or shares our bodies release the same chemical patterns as when we get bullied or neglected. The difference between chemical rewards on social media and those in real life is the frequency and intensity of the interactions.

Less Likely to Get Enough Sleep

Percentage of 8th through 12th-graders who get less than seven hours of sleep most nights[19]

† The gray area represents the time since the iPhone launched in 2007, marking the beginning of the mobile computing era.

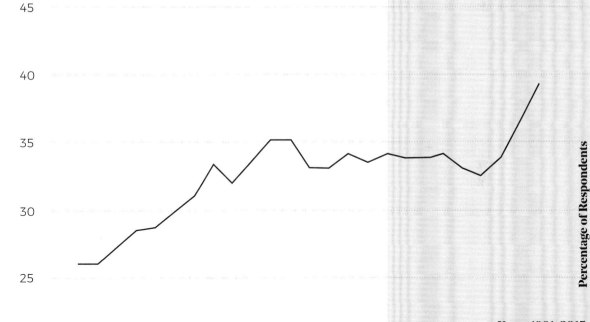

Percentage of Respondents

Years, 1991–2015

45
40
35
30
25

1991   1995   2000   2005   2010   2015

More Likely to Feel Lonely

Percentage of 8th through 12th-graders who agree or mostly agree with the statement "I often feel left out of things" or "A lot of times I feel lonely"[20]

■ Often feel left out

▨ Often Feel lonely

† The gray area represents the time since the iPhone launched in 2007, marking the beginning of the mobile computing era.

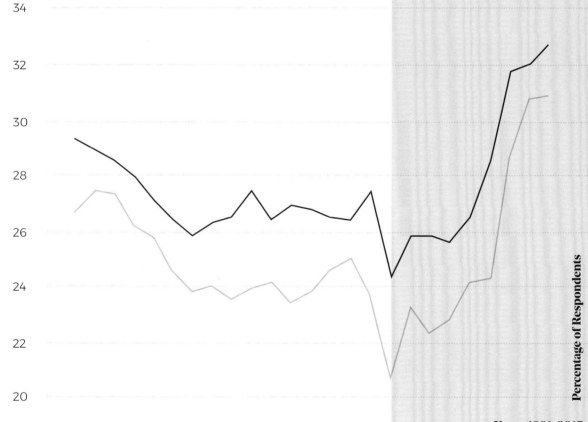

Percentage of Respondents

Years, 1991–2015

34

32

30

28

26

24

22

20

1991   1995   2000   2005   2010   2015

Losing Purpose

Percentage of 8th, 10th, and 12th-graders who are neutral, mostly agree, or agree that "My life is not useful," or "I do not enjoy life."[21]

■ "My life is not useful
▨ "I do not enjoy life"

† The gray area represents the time since the iPhone launched in 2007, marking the beginning of the mobile computing era.

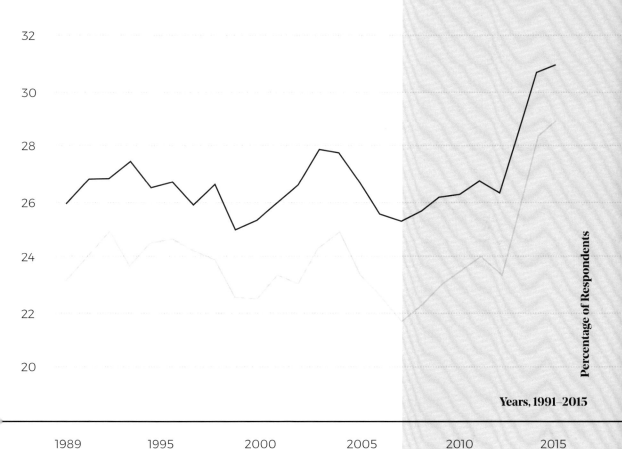

Percentage of Respondents

Years, 1991–2015

32
30
28
26
24
22
20

1989   1995   2000   2005   2010   2015

What used to take days, months, or years can now happen in hours, minutes, and, sometimes, seconds, depending on the person and the day. And unlike in the old days when people could hide in their room, there's no escaping. The pings come from around the world at the speed of the wireless connection. The only way to escape is to turn off our phones—something many people are incapable of doing. The compounding effects of all these interactions over time can create some nasty chemical reactions. Combine this with the fact that being engaged with a screen means being disengaged with everything else in near proximity, and what's clear to see is that these systems are isolating us from the world despite the fact that people are right in front of our face. We are evolving into a society full of people that are, as Sherry Turkle puts it in her book, alone together.

The compounding effects of social media use and extended periods of digital isolation are clearly reflected in the skyrocketing rates of loneliness and depression across society. In 2016, it was proven that consistent interaction with social media—whether that be liking things, sharing links, posting status updates, or all three—is proven to reduce physical and mental health, as well as overall life satisfaction.[22] This is painful to hear but it's even worse among kids and young adults. Between 2010 and 2015, depressive symptoms increased 33 percent, and were matched with suicide rates, which rose 31 percent among youth. These numbers were even higher among young women who experienced 58 percent more depressive symptoms, and committed suicide 65 percent more often.[23] These numbers are shocking, but they are just a small sampling of the various negative outcomes discovered by Jean Twenge, who writes about these effects in depth in her book, *iGen: Why Today's Super-Connected Kids Are Growing Up Less Rebellious, More Tolerant, Less Happy—and Completely Unprepared for Adulthood—and What That Means for the Rest of Us*.

Face It, We're Hooked

The compounding effects of physical pain, sleep deprivation, and addictive depression-driving apps that isolate us from each other are creating a perfect storm for what the people in Silicon Valley would call "disruption." But this disruption isn't focused on creating a new business model or technology. Instead, it's concentrated on disrupting human evolution and society, at large. We are now doing exactly what Nir Eyal set out to do—create products that change what we do and, by extension, who we are.

On March 28, 2017 I attended one of Eyal's workshops at the LinkedIn building in downtown San Francisco just to see what he had to say for himself after teaching the world how to systematically design addictions into products in order to drive revenue. Although I found the talk to be interesting, it was clearly Nir's way of doing PR in hopes of cleaning up the mess he's helped make. I respected his intentions but questioned his motivations. At the end of the talk, members of the audience were each allowed to stand up and ask one question, if they

Increased Symptoms of Depression

Depressive symptoms by sex, 8th, 10th, and 12th graders.[24]

■ Female
▦ Male

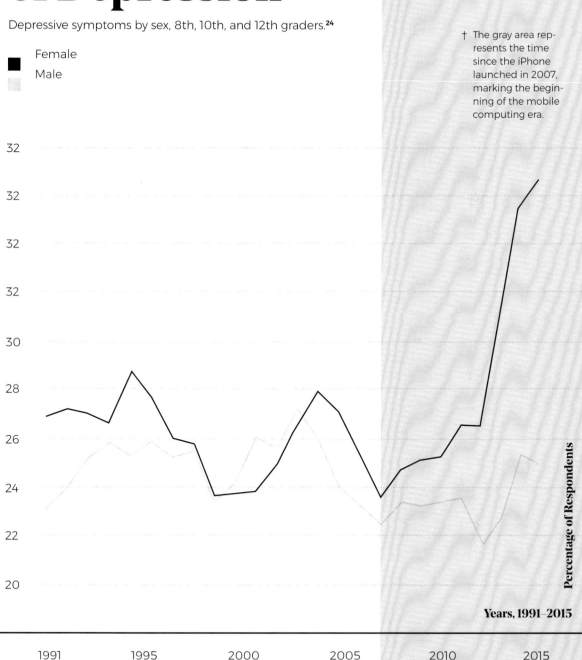

† The gray area represents the time since the iPhone launched in 2007, marking the beginning of the mobile computing era.

32

32

32

30

28

26

24

22

20

1991   1995   2000   2005   2010   2015

Percentage of Respondents

Years, 1991–2015

had one—and you better believe I had one. After waiting in line for several minutes I finally had the chance to ask Nir the question I had been dying to ask him for a while now. I grabbed the mic and said, "There's plenty of behavioral research that demonstrates the power the internet has to amplify and catalyze behaviors, but almost none of it says we have the capability to reverse behaviors through the internet. So my question to you is, do you truly believe we are capable of reversing this worldwide addiction through the devices that created it?"

Instead of directly answering my question, he pitched an app he had invested in to help addicts fight against their addiction. It was a good step, but I wasn't satisfied. I didn't ask the question to get a dismissive answer from the guy running the show. I refused to be quietly patted on the head and told to sit down. So, instead of going to sit down, I turned around to the woman directing the line and said, "Sorry, I know we're only supposed to ask one question, but I don't feel like my question got answered."

Without any confirmation that I could ask another question I turned to Mr. Eyal and said, "I appreciate the effort you're making to help those in need. It's honorable, and somebody needs to do it. But I feel like maybe I misstated my question... You know as well as the rest of the behavioral psychologists in this room that an addiction is not something that addicts just wake up and voluntarily decide to change one day. They don't look in the mirror and say, 'I have a problem, I should go download an app.'"

"Typically," I continued, "change only occurs because of a dramatic life event—their child saying 'Daddy, I think you're only happy when you're drunk,' the loss of a spouse who couldn't take it anymore, or a drunken car accident that kills somebody. This event strikes them so deeply that it awakens them to something greater and makes them want to change their life. So, to reiterate my question: Do you really believe that an app which users have to seek out can reverse an epidemic that now plagues our world?"

Despite his best efforts to answer the question he basically restated his previous answer. I gave him a nod in appreciation of his time, despite his inability to give a direct answer, because I knew that more than likely he doesn't know how to solve this problem better than anyone else. All we can do at this point in history is recognize that addictions have intentionally been installed at a global scale and are now becoming a threat to humanity. Unfortunately, unlike machines, this bug can't be solved with a software update.

CHAPTER 4

Herding the Sheeple

With the ability to shift attention towards whatever is most valuable to their bottom line, the general public has become a herd of mindless sheep fighting each other to get ahead instead of focusing on what really matters. Overcoming the breakdown of trustworthy communication will be our greatest challenge in this era of post-truth.

In any centralized information organization, a lot of power is given to a select group of individuals, putting them in a position to control the public conversation. The power granted to those in control is why we generally refer to the press as the Fourth Estate—an invisible political actor that influences the public in a way that often goes unrecognized. The downside of this centralized power is that any information that threatens the status quo is likely to be suppressed. But the upside is that everyone generally sees the same information, which enables a sense of stability within communities.

Maintaining this stability thousands of years ago in nomadic societies was not so difficult. Our understanding of the world was limited to where we lived and who we lived with, which allowed communities to align within themselves and their localities. But as the size of communities grew, the way we maintained informed communities changed with the invention of propaganda, or news, as we often call it today, which can be traced back to a time well before the printing press.[1] And as news turned into a global business, the ability to control large groups of people only grew larger, making the business of news exponentially more enticing.

From the first time television appeared in 1927 until 1976, when Ted Turner created the Cable News Network (CNN), there were only three national broadcast stations in the United States—the National Broadcasting Company (NBC), the American Broadcasting Company (ABC), and the government funded Public Broadcasting Station (PBS).[2] These companies controlled the public conversation in the States, and the same level of centralization was seen in many other countries around the world. The British Broadcasting Corporation (BBC), founded in 1927, and Independent Television (ITV), founded in 1955, dominated the United Kingdom until the 1980s and still hold a large share of the market today.[3,4] Similar centralized establishments have be seen in every developed and developing nation around the world. In fact, Brazil, China, and Germany, as well as many others, still operate under one main news source for the most part.

Although the people running these organizations may not be perfect, they had to be trained to get the job. This meant they had to understand the power they possessed, and how to wield it with integrity. This was coupled with lengthy editorial processes that acted as checks and balances within the system. As time went on and quality standards raised, journalistic objectivity was stressed in order to make sure the news reflected different points of view. And, in order to maintain integrity, news was often reported in a moderate, boring way in comparison to the way it is reported today. This all changed with the advent of the internet, social media, and citizen journalism, which have flipped the power structure of the information industry inside out.

Profit-Mediated Anarchy

Breaking up the power structures within the information industry has been part of the dream for the people building the internet since day one.[5] And today, thanks to those dreamers, the average person can be a publisher. Anyone can tell their story through blog posts, a series of photos posted online, a livestream of an event, or whatever it is that gets the message across. Services like YouTube, Facebook, Twitter, and many others have removed the need for professionally trained journalists and the average citizen has now become our source of news. "Citizen journalism," as this is called, has created an era where anyone can have a voice and has the opportunity to be heard equally. This has led to more social progress in the past 20 to 30 years than we have seen in the history of humanity. But while the Internet has given a voice to many who otherwise may have been suppressed or marginalized, it has also enabled the demons of humanity to assemble and generate incredible momentum. Larry Page and Sergey Brin clearly outlined these dangers in their 1998 PhD thesis, "The Anatomy of a Large-Scale Hypertextual Web Search Engine":

> *Another big difference between the web and traditional well controlled collections is that there is virtually no control over what people can put on the web. Couple this flexibility to publish anything with the enormous influence of search engines to route traffic and companies which deliberately manipulating search engines for profit become a serious problem* [sic]. *This problem that has not been addressed in traditional closed information retrieval systems. Also, it is interesting to note that metadata efforts have largely failed with web search engines, because any text on the page which is not directly represented to the user is abused to manipulate search engines. There are even numerous companies which specialize in manipulating search engines for profit.*[6]

While McCarthy and other internet founders desired a decentralized source of information where everyone could have their own opinion, they also feared monopolists and other morally corrupt individuals interested in wielding power. And if we examine the current state of the internet, it's easy to see this is exactly what has happened.[7] Instead of being safe and democratic, the internet has become anarchy, a rush to obtain the most followers, which creates trust, which then increases the ability to gain more followers, and more trust—a vicious circle that catalyzes both influence and trust to the point that an individual's authority becomes unquestionable and their ability to influence almost certain. Add to this the ability to influence people around the world from the push of a button, and what we've enabled is a centralized form of power like nothing we've ever seen before.

Fake News

The scapegoat of modern media is fake news. It's the first thing every-one points to when trying to determine what makes the internet so dangerous. However, fake news has been around forever. Look at your local grocery store checkout aisle. See that fresh new copy of *National Enquirer*? That's fake news, but you know that already. On TV there's Fox News and MSNBC. And on the internet the list could go on forever. It's not that all of the content these organizations create is fake, but that there is enough fake news that people on opposing sides feel incapa-ble of trusting these sources—a sentiment that is not entirely wrong.

Reporters for these organizations ask questions of their audi-ence and expect them to report back with the answers. They report stories based on speculation and throw in sound-bites without context to make things seem worse than they are. And, more generally, they promote the news in a way that drives fear and outrage. But they're not alone. This is the business of news on the internet, driven by algo-rithms that thrive on sensationalized fearmongering—a problem that now plagues our world thanks to the way the internet makes money.

Although news organizations get the heat for what's happen-ing, it also has to be recognized that they've been forced into this. From the year 2000 to 2015 the ad revenue from newspaper subscriptions dropped from nearly $50 billion to less than $20 billion, forcing tradi-tional news media companies to adapt their strategies to a 24-hour news cycle based on fearmongering and sensationalism.[8] This means due process, journalistic integrity, and moderate opinions have gone out the window in favor of whatever will get readers to click. And now, because of this, we have more fake news than ever before—including information from the people we used to be able to trust, which is what makes it so dangerous. Even so, the problem isn't fake news, it's that there's so much more of it because information can be created and spread cheaper, faster, and, in general, more easily than at any other point in history. The path of least resistance has been paved, and we now have more "experts" than ever before, many of whom have no idea what it means to maintain journalistic integrity, professional objectivity, or any of the other traditional standards.

With the exponential increase in internet "experts," the amount of information coming from the internet is now magnitudes greater than that which came from any prior organization or group of organi-zations, and it's moving at the speed of thought. This has divided the public by creating dense walls of polarizing media and information that shapes people's minds and makes it impossible for them to main-tain a broader perspective on the world. These walls are invisible, but the dangers are real and they have created a painful state of cognitive overload in the public's mind. The people are drowning, and they're beginning to give up.

Echo Chambers

Another problem many people overlook is that with algorithms distributing content based on our clicks, it's easy to get lost in a rabbit hole of information personally designed just for them. And the more people click, the more personalized the service gets. But there is an inherent flaw to this idea. If, for example, this idea were applied to a highway and on which there was an accident that everyone kept looking at, the internet would think we wanted more accidents.

Being defined by likes, follows, and other forms of subscription makes people feel like they're getting what they want, but in reality it is actually putting them in a dangerous position. Too much personalization leads to self-validation feedback loops, which results in people seeing only what they want to see and liking what they're given to look at, which then reinforces the algorithms and pushes people to even greater extremes. But with all of this happening behind the scenes, echo chambers form invisibly, leading to a dangerous disconnect from the rest of the world that is nearly impossible to recognize because when all people see is one side of a problem or issue this becomes all they are capable of understanding. And now, because of this, we have entire populations of people living in their own personal realities.

Despite the hyper-personalized information each of us experiences individually, the internet has also brought us back to an era of centralized information. But this time it is global, not national, and the centralization has more to do with the way information is distributed than with what's being created. The companies running these distribution systems have, through our likes, follows, shares, and other forms of subscription, been given the right to define what is and isn't meaningful in our lives. But their methods, based on fear-mongering sensationalism, have created a world of hyperpersonalized realities where it is now easier to fool people than to convince them they've been fooled.

Surveillance and Targeting

The convergence of Big Data and entire populations blind to the world outside of their own echo chamber, has made manipulating the world easier than ever before. And, according to recent reports, upwards of 95 percent of companies around the world do this.[10] It's called targeted marketing, and it has been strategically used to shape the world for at least the last half-century. The Nixon presidential campaign team was one of the first groups to leverage it in a meaningful way. They did this by creating focus groups from around the country to understand the desires and needs of people in specific regions better. After analyzing an audience at an individual level, the team could then create messages that resonated personally with individuals, and send these specially crafted enticements through direct mail. At the time, this meant neighbors could be sent different pieces of mail, each receiving a message that resonated better with them personally. As Cathy O'Neil put it in her book *Weapons of Math Destruction*: "Direct mail was microtargeting on training wheels."[11]

1   The splatter of data, visualized above, was discovered by William J. Brady and his team at New York University. It shows how tweets containing moral or emotional content were shared during the 2016 United States election—clear evidence of how deeply people are embedded in echo chambers.[9]

Today targeted marketing plays a powerful role in our world, but it has become much more insidious. Recent reports have shown that in 2016 alone, this type of behavior was used to influence the elections of at least 18 different countries.[12] The countries listed in the reports are some of the most "developed" nations in the world, yet they're all imploding in on themselves. But why? Well, consider what these countries all have in common. Upon a quick survey we can determine that they don't have similar governments, languages, currencies, or cultures, but what they do all have is the invisible hand of the internet.

As we'll discuss further in chapter five, the data analytics teams at Facebook, Google, and every other major organization on the internet know just about everything that's happening, both on their platform and in people's lives. These companies make their money by understanding data and helping others target individuals. It's a fact they are incapable of denying. This is how Twitter made nearly $4 billion, Facebook made nearly $40 billion, and Google made more than $96 billion in 2017.[13][14][15]

The thing is, no matter how much anyone wants to get angry at them, this is all legal. Companies can legally buy and sell data from each other.[16] This means that if you do something on Facebook, Google, or any other platform that collects data, any company with the money and desire to purchase that data can do so legally. Since the 2016 election this has become an inflammatory topic thanks to partisan "experts" who are more interested in defending their party's decisions than actually discussing the issue, but the reality of the story is that the data Cambridge Analytica was pulling from Facebook only represents an infinitely small percentage of the data out there. Millions of other firms around the world are doing the same thing every day, but they're doing it silently by buying and selling data the public is not even aware of. And more than likely they're doing it in ways that are nearly impossible to track.

Does this mean I agree with what Cambridge Analytica did or the role they played in recent elections? No. But, disagreement is part of democracy. In fact, the freedom to disagree is what makes the democratic process so powerful. However, in any zero-sum game one side will always end up "losing" to the other, which almost always ends with that side being upset. The same could have been said when Obama won, and his campaign's social media strategy was celebrated.[17] The question that anyone who is angry about the previous elections needs to ask themselves is this: *If the side I voted for had won, would I have been as mad?* Better yet: *Would I have cared about or been aware of the potential for interference?* This isn't to make a statement about whether what happened should be considered any more or less malicious, or to say that there was no interference involved. Those are issues for someone else's book and are still, at this point, unknown. The point of this question is, however, to recognize that the opposing parties in these elections could have done the same, they just didn't—or at least not as well. Was this work ethical or should it be considered to be? That's up for debate. But was it legal and is it done by millions of organizations around the globe on a daily basis? Yes, absolutely.

2 If Facebook charged everyone $2 a month, they would make more money than they did in 2017 and wouldn't have to serve a single ad.

Large Hadron Collider's annual data output
15,360 TB

US Census Bureau data
3,789 TB

National Climatic Data Center databa
6,144 TB

Google's Search Index
97,656 TB

Nasdaq stock market database
3,072 TB

Videos uploaded to YouTube per year
15,000 TB

Kaiser Permanente's digital health records
30,720 TB

Content uploaded to Faceboo
182,500 TB

Tweets sent in 2012
19 TB

Business emails sent per year
2,986,100 TB

Library of Congress' digital collection
5,120 TB

The Size of Data

ach year

† All data from 2012[18]

The Last Days of Reality

After reading all of this, we're now back at the first question we started with: *What is the most dangerous part of the internet?* If it isn't the mass amounts of fake news, the echo chambers, or the targeted marketing done so specifically that it feels like we're being spied on, then what is it? While these aspects of the internet are menacing, the most dangerous is that these elements have all aligned, forming an invisible, worldwide system of profit-mediated anarchy controlled only by those literate enough to understand what is going on.

Facebook and other companies have claimed that they haven't brought in any humans to curate content because it would cause bias.[19] This sounds reasonable, but at the end of the day this is just rhetoric from public relations teams looking to satisfy the public. Meanwhile, on the back end every system is an algorithm acting as an automated, invisible form of logical government that has ultimately been programmed by a small group of humans to do what a team of humans would do otherwise. And yet, even the idea of algorithms built to govern the internet is not the problem. Every space needs rules, and with a space as large as the internet, an automated form of government is the only possible way to maintain those rules.

The real problem is that the rules governing these algorithms, which ultimately govern the internet, are being defined by an incredibly small, homogeneous group of individuals who hold an incredible amount of power and can hide behind a black box inside publicly traded companies making billions off the damage. While it would be ideal to be able to trust these companies to create meaningful communication platforms that help societies around the globe flourish, it isn't realistic to expect that kind of behavior from a publicly traded corporation trying to perpetually increase revenue.

So while we're distracted by our inability to decipher truth from fiction, unwilling to work together to get ourselves out of this mess, these companies are going to continue using their power to make billions as they shape us in ways most of us are incapable of recognizing. And, if we allow it, things will only get worse as the digital era unfolds and the proper precautions are not put into place. In fact, recent developments in research labs have discovered ways to make videos that can blend two individuals' voices and faces together to make fake videos that look so real that people won't be able to tell the difference. Imagine, for example, a video of Donald Trump talking about how much he loves Hillary Clinton and not being able to tell if it's real or not. What this means is that we're entering an era of post-truth, where we will struggle to tell truth from fiction. Left to profit-driven algorithms, these dangers will surely consume us.

CHAPTER 5

Social Engineering 101

Distracted by the extremism in the news, Silicon Valley can leverage its power to silently manipulate the market in ways the public is incapable of seeing. Left alone, the market will be swallowed by these companies without any visible signs of destruction until it's too late.

In 1949 George Orwell published a book titled *1984*, in which, the main character lives in a world that is perpetually monitored and analyzed for variances against the norm—or at least the "norm" according to the Thought Police, the people who governed the world. In Orwell's story, this perpetual surveillance creates a state of paranoia and learned helplessness among the older citizens who remember what life was like before all the surveillance eliminated their freedoms—a life that the younger generation is incapable of understanding because they were born into the new world and know no different. But for those that do remember, any talk about what life used to be—even just the thought of it—could be considered a crime:

> *Whether he wrote DOWN WITH BIG BROTHER, or whether he refrained from writing it, made no difference... The Thought Police would get him just the same. He had committed—would still have committed, even if he had never set pen to paper—the essential crime that contained all others in itself. Thoughtcrime, they called it...was not a thing that could be concealed forever. You might dodge successfully for a while, even for years, but sooner or later they were bound to get you... In the vast majority of cases there was no trial, no report of the arrest. People simply disappeared, always during the night. Your name was removed from the registers, every record of everything you had ever done was wiped out, your one-time existence was denied and then forgotten. You were abolished, annihilated: vaporized was the usual word.*[1]

Though Orwell's *1984* is fiction, if we consider what's going on in the world today, it's easy to see that we're not far off from allowing this to occur. Trying to turn down the products and services created by Big Tech has become nearly impossible. Many, in fact, haven't even needed to advertise until the past few years because of their dominance, and even then the ads are more likely to be a PR statement demonstrating why they should be trusted than a traditional product advertisement. Unfortunately, such dominance has allowed Big Tech to grow powerful beyond recognition—to the point they're now capable of directing the market—and we're only just beginning to see the effects. Left alone, Big Tech's power will only become greater.

If we want to see where we might be headed, we can look to China where CCTV, a state-run broadcast company, has set up a system of 170 million cameras and police wear sunglasses with facial recognition software to help them police the population more efficiently. To demonstrate the power of this system, BBC reporter John Sudworth visited Guiyang, China, a city with a population of more than 4 million people, to test it out, and within less than seven minutes of being dropped off he was handcuffed and arrested.[2][3] This, on top of China's new Social Credit Score which tracks everything citizens do—from the things they say on social media, to the places they shop, to the way they travel, and more—doesn't sound so far off from Orwell's novel does it?[4] Don't worry though, the Chinese government has already told everyone, "If you have nothing to hide, you have nothing to fear."[5]

Although this is the situation in China, an authoritarian nation known for its surveillance and information control practices, the situation in the most developed nations is not much different. We already have camera phones with facial recognition; millions of people rate each other on apps every day; and, on top of these voluntary behaviors, the companies running the show have incredible power to define the information we see. Our world is only a few flipped switches from reflecting what's happening in China.

Apple's App Store Control

A market, in the traditional sense, is a place where people gather to buy and sell goods. Nobody owns the market, and in order to participate business owners had to have something to offer. Often this inventory was bought at a bulk price and then sold to consumers at higher prices, or "market value." This additional cost on trade was how businesses made profit, in exchange for the risk they took when they purchased the inventory without knowing whether it would sell or not. Apple has flipped this model upside down by creating a marketplace that they own and govern as they see fit, and it's one of the most brilliant business strategies the world has ever seen.

Instead of purchasing products up-front, Apple gets app developers to build and market the products for them, then makes 30 percent off all sales in the App Store, including in-app purchases that occur after the consumer has already downloaded the app. This means if you download an app, then leave the App Store and buy something directly through the app, Apple still makes money off that transaction. This would be like if someone bought Legos at Target then bought accessories through Lego.com and 30 percent of the profits made on Lego.com had to be paid to Target. Oh, and if developers do any marketing for their app they're also marketing for the App Store because people have to go to the App Store to download the app. It's a win-win for Apple, who doesn't have to build the product, market it, or do much of anything, and still makes billions—almost $8 billion in 2016 alone.[6]

The existence of the App Store, like all great internet platforms, is a chicken-and-egg dilemma. Without the App Store, developers would have trouble marketing and distributing their products, but without developers, the App Store wouldn't have a wide variety of apps to sell and would likely cease to exist. Despite this, owning the App Store allows Apple to govern the territory as it sees fit, and the company is aware of this power, which is clearly demonstrated in the Introduction section of the App Store Review Guidelines, which states: We will reject apps for any content or behavior that we believe is over the line. What line, you ask? Well, as a Supreme Court Justice once said, "I'll know it when I see it."[7]

Amazon's Retail Control

Similar to Apple's App Store, Amazon has established an unsurpassed control of the retail market. In 2014, at Book Expo America, the American Booksellers Association honored best-selling author James Patterson in recognition of the $1 million pledge he had made to support independent booksellers across the United States. At the ceremony Patterson gave a speech addressing why he felt so passionate about making the donation and emphasized the power Amazon has over not only booksellers, but retailers worldwide:

> " *Right now bookstores, libraries, authors, publishers, and books themselves are caught in the crossfire of an economic war between publishers and online providers. To be a teeny, tiny bit more specific, Amazon seems to be out to control shopping in this country. This will ultimately have an effect on every grocery- and department-store chain, on every big-box store, and ultimately it will put thousands of Mom-and-Pop stores out of business. It just will, and I don't see anybody writing about it, but that certainly sounds like the beginning of a monopoly to me.*[8]

Patterson's speech was directed at the way Amazon had been treating him and his team at Hachette Book Group. The battle between the two companies led to Amazon removing Hachette titles from the Kindle store, ranking Hachette products lower in search results, and intentionally delaying the shipment of Hachette hardbound books in order to anger consumers.[9] Amazon can do this because, like Apple, it owns the store and can run it as it pleases. But, of course, Amazon hasn't stopped at book sales or limited itself to the United States. As Patterson pointed out, the company's goal is not to dominate book sales, clothing, food, or any specific retail vertical but to monopolize the global market, in general. This sounds unreal, and traditionally would take a lifetime, if not longer, to accomplish, but in the age of the internet this is completely realistic. The company's acquisition of Whole Foods in 2017 reflects how quickly this may happen.

In June 2017 Amazon bought Whole Foods for $13 billion. While this price tag sounds outrageous, it was actually quite the deal. Instead of constructing new buildings across the country like traditional retailers must do, a process that takes a lot of time, money, and risk, Amazon just bought 460 stores instead.[10] But not just any stores, stores whose demographic matches Amazon's target demographic for Amazon Echo users. Now, instead of using Instacart, GrubHub, or whatever other delivery service these people might use otherwise, they can operate directly through Amazon, there's no need for a middleman.

As an added bonus Amazon can also leverage the physical locations to market their Echo products, which means Whole Foods is no longer just a food market but also an electronics store. Why would Amazon do this? Because people buy more through Echo than they do when they shop in stores.[11] This is due to a common psychological trick called the cashless effect, in which people spend more if they don't

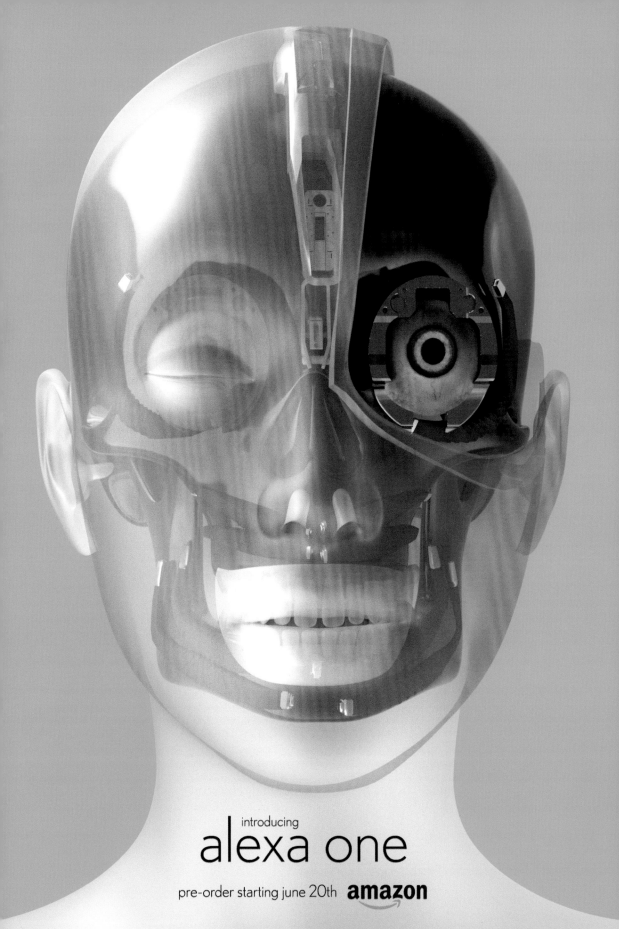

introducing
alexa one

pre-order starting june 20th   amazon

see the money—something retailers have known about since the beginning of credit cards.[12]

1 What happens when the majority of retail occurs through voice activated devices and results default to Amazon products, as researchers have discovered is already happening?[13]

Furthermore, Amazon has also created the Echo Wand, an evolution of the Amazon Dash button. Now, instead of having a button for each product, people can use a barcode scanner in their home to make their grocery list and push one button to complete the order, which then gets delivered directly to their door. This, in combination with the upcoming smart refrigerators and pantries of the future, will be how grocery purchasing gets automated. It will be so easy people won't even have to think about it anymore, which is exactly what Amazon wants—a subscription service for our basic needs.

In addition to automating purchases outside the store, Amazon is also working on automating purchases in the store. The future of retail can be seen in Amazon Go stores, which allow customers to walk in, grab what they want, and walk out without ever interacting with a cashier or payment platform, allowing Amazon to eliminate the cost of human employees. When this happens, Amazon will be able to cut prices so low that no traditional retailer will be able to compete. Even if thousands of items are initially offered at prices that cause Amazon to lose money, the company has the capital to take the hit, mark these items as "loss leaders" on the books, and force other stores to give in or give up. If traditional retailers hope to compete with Amazon on price, that money will either have to come from their bottom lines or their employees' pockets—neither of which they can afford.

On top of all of this Amazon has also begun automating every point along the supply chain. On February 9th, 2018 the company announced that it is officially moving forward on its project, "Shipping with Amazon" which will use the fleets of delivery vehicles, Boeing 767s, and cargo ships they've previously purchased to eliminate the need for outside services.[14][15] Amazon has also patented infrastructure for connected roadways, which we can safely assume will be used to guide fleets of automated delivery vehicles once the technology is ready.[16] The company has also purchased an intelligent doorbell company, called Ring, which will surely be used to check packages in and out while also keeping them safe from burglars.[17]

It's clear that Amazon is focused on eliminating any humans unnecessary to its operations, which threatens millions of jobs worldwide. But Amazon can do this because the company has the technical capabilities to create these systems, the influence to define the market, and the capital to lower costs in a way that will strangle out any competition—something Jeff Bezos clearly outlined in a letter to shareholders when he noted that Amazon has seen both great successes and "billions of dollars' worth of failures along the way." Making billions of dollars worth of mistakes along the path to charging customers more money is something only a company like Amazon can get away with, because of their powerful reach.[18][19] Such influence, paired with its technological dominance, is how Amazon will monopolize retail worldwide.

Facebook's Emotional Control

Owning a marketplace like Apple and Amazon defers an incredible amount of power, but owning a social media platform has distinct benefits that no commercial marketplace will ever see. To start, consider that over 2.1 billion people, nearly one-third of the world, have a Facebook account, and the average user dedicates 39 minutes a day to the platform. Combined with the other major platforms Facebook owns, including Instagram and WhatsApp, the time people spend with the brand rises to well over 50 minutes per day. This means people are spending more time with Facebook than any other activities apart from work, sleep, or interactions with family.[20]

All of this data gives Facebook unprecedented insights into people's habits and interests. To understand just how well it knows users, Facebook commissioned a study in 2014 to learn more. After analyzing results from over 86,000 volunteer participants, the company concluded that with 10 likes, it is capable of judging people better than their co-workers. After 150 likes it knows people better than their family members. And after 300 likes the company can predict people's opinions and desires better than their significant others.[21]

Around the same time Facebook commissioned another study titled "Experimental Evidence of Massive-Scale Emotional Contagion through Social Networks." The study was done by research scientists at Facebook, who intentionally manipulated users news feeds to get a better understanding of Facebook's potential to emotionally manipulate the world. This means Facebook manipulated people to understand how well they could manipulate people without them being aware of it. The study is prefaced by the following statement:

> Questions have been raised about the principles of informed consent and opportunity to opt out in connection with the research in this paper. The authors noted in their paper, "[The work] was consistent with Facebook's Data Use Policy, to which all users agree prior to creating an account on Facebook, constituting informed consent for this research." When the authors prepared their paper for publication in PNAS, they stated that: "Because this experiment was conducted by Facebook, Inc. for internal purposes, the Cornell University IRB [Institutional Review Board] determined that the project did not fall under Cornell's Human Research Protection Program." This statement has since been confirmed by Cornell University.

> Obtaining informed consent and allowing participants to opt out are best practice in most instances under the US Department of Health and Human Services Policy for the Protection of Human Research Subjects (the "Common Rule"). Adherence to the Common Rule is PNAS policy, but as a private company Facebook was under no obligation to conform to the provisions of the Common Rule when it collected the data used by the authors, and the Common Rule does not preclude their use of the data. Based on the information provided by the authors, PNAS editors deemed it appropriate to publish the

2 How many posts have you liked on Facebook since the day you signed up?

3 Facebook received severe public backlash for treating the world as a test lab, but legally it did nothing wrong. This behavior is covered in the terms and services agreement that everyone agreed to when they signed up for Facebook.

4 If you hear anyone talk about people being uploaded to the cloud, it isn't about literally plugging people into a machine and sucking them up into a server, it's about the fact that almost all of our behavior since the day we were born has been recorded—a fact that is especially true of younger generations, whose parents have been documenting their lives since the day they were born.

paper. It is nevertheless a matter of concern that the collection of the data by Facebook may have involved practices that were not fully consistent with the principles of obtaining informed consent and allowing participants to opt out.[22]

While likes tell Facebook a lot, the reactions feature that launched in 2016 gives the company an even better window into people's souls. Facebook now knows people's emotions so well it has even begun to sell this data to marketers who want to target people based on their emotional states. In May of 2017 Facebook was caught selling emotional data to marketers that would allow them to target 13- and 14-year-olds in their most desperate moments of need.[23] Moral? No. Profitable? Yes, very. But don't think Facebook is about to stop there. The company took another step towards targeting children by releasing Messenger Kids at the end of 2017 to onboard 6- to 12-year-olds. The app is marketed as, "A way for parents to take control of their kids internet usage," but don't be so certain of their claim to good intentions.[24] The app will not only give Facebook more users, but will also allow it to get an in-depth view into what this age group is interested in, which will allow Facebook to strategically market to this population.[25]

We've seen manipulative targeting of children before by Big Tobacco when Joe Camel was created. Just like Big Tobacco, Facebook has saturated its mature market and is now attempting to make inroads into the infancy of its global audience, which it historically hasn't had access to for legal reasons. The difference is that Big Tobacco didn't have God-like knowledge about its consumers and competition. Eventually the practices of Big Tobacco were deemed illegal by the Federal Trade Commission (FTC) for "appealing to many children and adolescents under 18, induced many young people to begin smoking or to continue smoking cigarettes."[26] Will Facebook and other companies be held to the same standard?

Google's Knowledge Control

While the power that resides within Facebook's ecosystem is impressive, the walled garden of Google, and its parent company Alphabet, is magnitudes larger. The company's influence is so wide-reaching that most people don't even realize they're using Alphabet products or that competing products exist. For example, everyone knows Google is the number one search engine in the world, but what about number two? I bet you'd struggle to guess what company is number two. In fact, you may even need to Google it. But instead of doing that let me just tell you that the number two search engine in the world is YouTube—which also happens to be owned by Alphabet. Between the two, the company processes more than 3.5 billion queries every day on Search, and on YouTube people watch nearly 5 billion videos every day.[27][28]

Alphabet also owns Gmail, which represents decades of all the most intimate thoughts, conversations, and network interactions of billions of people around the globe; Android, the operating system for more than 86 percent of mobile devices worldwide as of the first quar-

ter of 2017, which gives the company access to nearly every interaction on users' devices if they want it; and Maps, which perpetually tracks the location of billions of people every second of the day.[29] But despite how astounding the numbers for these five platforms are, these are just a small handful of the nearly 500 tools that Google has created, many of which people don't even think or know about.[30]

With these tools, Google has access to our most intimate thoughts and desires. For example: Need to know what's going on in the world today? Google. Need to know what that acronym your boss just said means? Google. Have a question about something you're too embarrassed to ask anyone else? Google. Am I right? With all of this data, Google knows more about you than you know about yourself. The company is as close as it gets to being God. But don't take it from me, take if from Google's former CEO, Eric Schmidt himself, who was quoted In a 2010 interview with *Atlantic* magazine saying, "We know where you are. We know where you've been. We can more or less know what you're thinking about."[31] And then consider the fact that this quote is from 2010, nearly a decade ago.

As Alphabet pursues projects like Sidewalk Labs, which is focused on connected cities; Waymo, which is focused on automated driving; and Verily, which is focused on digital health; among others that exist in the physical world, the company will uncover and organize data previously unavailable through traditional software. Combined with the data they've been collecting from Search and every other product they own for nearly two decades now, this data will give Alphabet an unprecedented level of intelligence on the world.

Why is this is so troubling? Well consider, for example, that Search results are what Google believes we want, when it believes we want it and that they're capable of manipulating their algorithms the same way every other company discussed in this chapter can and has. But unlike an app store or even a retail marketplace, the information we receive from searches dictates the way we understand reality. This puts the company in a powerful position to define reality. And when that reality is based on advertising revenue it can become dangerous. In their 1998 PhD thesis, "The Anatomy of a Large-Scale Hypertextual Web Search Engine," Sergey Brin and Larry Page made note of these dangers long before Search became the source of power it is today:

> **"** *Currently, the predominant business model for commercial search engines is advertising. The goals of the advertising business model do not always correspond to providing quality search to users. For example, in our prototype search engine one of the top results for cellular phone is "The Effect of Cellular Phone Use Upon Driver Attention," a study which explains in great detail the distractions and risk associated with conversing on a cell phone while driving. This search result came up first because of its high importance as judged by the PageRank algorithm, an approximation of citation importance on the web. It is clear that a search engine which was taking money for showing cellular phone ads would have difficulty justifying the page that our system returned to its paying advertisers. For this type of reason and*

historical experience with other media, we expect that advertising funded search engines will be inherently biased towards the advertisers and away from the needs of the consumers.[32]

And in fact, the company has recently been sued for this exact kind of behavior by a company called Foundem, from the European Union, who took Google to court alleging that the company was discriminating against search results to serve its own interests.[33] The case resulted in a $2.7 billion fine that, as of September 2017, Google is attempting to appeal, but this isn't the first time they've been investigated.[34] The company was also investigated for antitrust behavior and search discrimination in the United States by the FTC.[35] That investigation resulted in a null decision, but even the FTC admitted that they were not entirely sure what they were looking for, just as Page and Brin had predicted:[36]

" *Since it is very difficult even for experts to evaluate search engines, search engine bias is particularly insidious. A good example was OpenText, which was reported to be selling companies the right to be listed at the top of the search results for particular queries. This type of bias is much more insidious than advertising, because it is not clear who "deserves" to be there, and who is willing to pay money to be listed. This business model resulted in an uproar, and OpenText has ceased to be a viable search engine. But less blatant bias are likely to be tolerated by the market. For example, a search engine could add a small factor to search results from "friendly" companies, and subtract a factor from results from competitors. This type of bias is very difficult to detect but could still have a significant effect on the market.*[37]

5 To explore this concept, search anything on Google then have a friend type the exact same words and compare results. If even one result is different (even if the results are just in different order), it is proof that Google makes results unique to each individual's online behavior.

Try this with several different people.

The bullet point that tends to get glossed over when discussing these cases is that these two investigations were only initiated because Google was bothering organizations that had the money, power, and knowledge to fight back. Many of the small- and medium-sized companies Alphabet is swallowing up don't have the opportunity to fight back. And with all the power that resides within the company, the likelihood of this situation getting better without regulation is low, especially considering what's being done to prepare for the future. In the fall of 2017, after the investigation in the EU, Alphabet decided to reorganize their operations under the holding company XXVI. This enabled Google, the holding company's main source of revenue, to be labeled as an LLC instead of a public corporation. This means Google now has only one investor it needs to report to—Alphabet. And with Alphabet as its only investor, Google is no longer obligated to publicly disclose any information about its earnings from Search, if it doesn't want to.[38] This is just the tip of the iceberg though.

Other things to be aware of include the fact that Google's employment contracts keep employees silent—even if they feel like what they're doing might be considered a crime; since 2012 the company has become one of the top lobbying organizations in the United States; and in late 2017 Eric Schmidt stepped down as CEO to straddle both the public and private sector as a technical advisor to both Google and the Department of Defense.[39][40][41][42] Despite how important these facts are almost none of them have become mainstream news.

With each step the company takes to escape public scrutiny and protect its privacy it appears to be avoiding the exact behavior it proposes the public willfully submit to—complete transparency through continuous surveillance and oversight. This is ironic considering Eric Schmidt once stated, "If you have something that you don't want anyone to know, maybe you shouldn't be doing it in the first place"—a statement that is eerily similar to that of the Chinese government when it openly admitted to using data to police society.[43] If we continue down this path of unquestioned faith in one of the largest publicly traded intelligence agencies in the history of the world there is incredible risk for manipulation at a scale that's never been seen before, which means the time to start asking questions is now.

Hand Over the Keys

In a 2014 TED Talk titled "Why Privacy Matters," Glenn Greenwald, who first reported the National Security Administration (NSA) leak from Edward Snowden, spoke about the state of the internet and surveillance. In his talk, he compared what's happening with today's lack of privacy to an 18th-century invention called the Panopticon—a circular prison with a single guard tower in the center, invented by social reformer Jeremy Bentham as a humane alternative to the cruel punishments occurring in the early 1800s.[44] Bentham's hypothesis was that if troublemakers were contained within a circular prison that had a tower in the center where guards could watch without being seen, prisoners would be forced to be good because they wouldn't know whether or not they were being monitored. Greenwald then discussed his concern that this is an exact reflection of what's happening within governments and large tech corporations worldwide.

Such surveillance systems are not only a threat to freedom of speech but also a threat to our right to organize, our freedom of self-expression, and, ultimately, our freedom of thought, in general. These developments amount to what could be considered a mental warfare that is silently swallowing entire populations around the globe at the speed of thought, largely without any public awareness. If we allow such activities to continue without question—if we allow these systems to direct our behaviors through their algorithms as they already do—we will see a global indoctrination of the people, by machine. But it will be done so well it will feel like it was our choice. And in a sense, it will be.

6 Google's motto "do no evil," doesn't necessarily mean "do good," it just means don't be evil. But what exactly does evil mean? What are the objective measures of evil? And where does that line get drawn when it comes time to increase key performance metrics for annual reviews?

CHAPTER 6

Creating Model Societies

With complete control of the market, companies can run things without question. Because of this, we've entered an era that—due to the abuses we are seeing—is eerily similar to the first Industrial Revolution. The difference now is that the abuse is not physically bound. It is, however, just as bad, if not worse, because of how hard it is to recognize.

At the beginning of the Industrial Revolution, work was dangerous for employees. People were forced to work long hours and there were no benefit packages, child labor laws, or regulations surrounding the cleanliness or stability of the work environment. More importantly, there was no way to oppose these conditions and challenge the people in charge. The impact of these conditions is reflected in books like *The Condition of the Working Class in England*, in which author Friedrich Engels wrote that factories were, "so dirty that the inhabitants can pass into and out of the court only by passing through foul pools of stagnant urine and excrement." Other stories like that of the Lowell mill girls who were forced to work from sunup to sundown every day, averaging 73 hours a week, spoke to the long hours people were forced into. And later, incidents like the Triangle Shirtwaist Factory fire of 1911, where 145 workers died because they were locked in while the building burned to the ground demonstrated the ruthlessness.[1][2][3] Eventually labor unions, standardized working hours, sanitation laws, worker benefit packages, and other safety measures were established, but what most people forget is what workers had to go through before these changes were made.

Today things may be different, but there's no denying there are still plenty of ruthless employers and toxic work environments. Companies offer unlimited vacation but then develop a culture where nobody ever leaves. They pay for employees' phones with the unwritten expectation that responses are required at any time of the day. They provide lunch because it means workers spending less time away from their desks. Some companies, like Three Square Market in Wisconsin, even embed chips into willing employees to help "make navigating the office easier" than if they had to carry a badge around.[4] This is the story of working in tech. The modern workforce may not be physically constrained like previous generations, but it is mentally constrained more than ever before. And on top of it, workers have again been placed in a position where they are incapable of fighting back.

Although people fought hard for fair representation after what happened during the Industrial Revolution, today labor unions are relatively non-existent. Contracts are written to suppress or eliminate the opportunity for class-action lawsuits. Standard eight-hour work days have become a luxury archived in history books, in exchange for a culture that is "always on." And traditional employee benefit packages offered to workers of yesteryear have disappeared for the majority of people who now are forced to work in the contract-based "gig economy." Yet these conditions go unquestioned due to the polarization of wealth and power that enables tech companies to squash any bad PR or fight any case brought to court. Further, although a large majority of this has to do with the decisions made by the people at the top, algorithms and automation have also been silently creeping into the equation for years now, which only makes everything harder to fight.

Today, full-time workers at the biggest tech companies may have more visible benefits than the majority of working people, and definitely more than their predecessors, but modern work at these companies is more like life in a velvet-covered jail cell than it appears

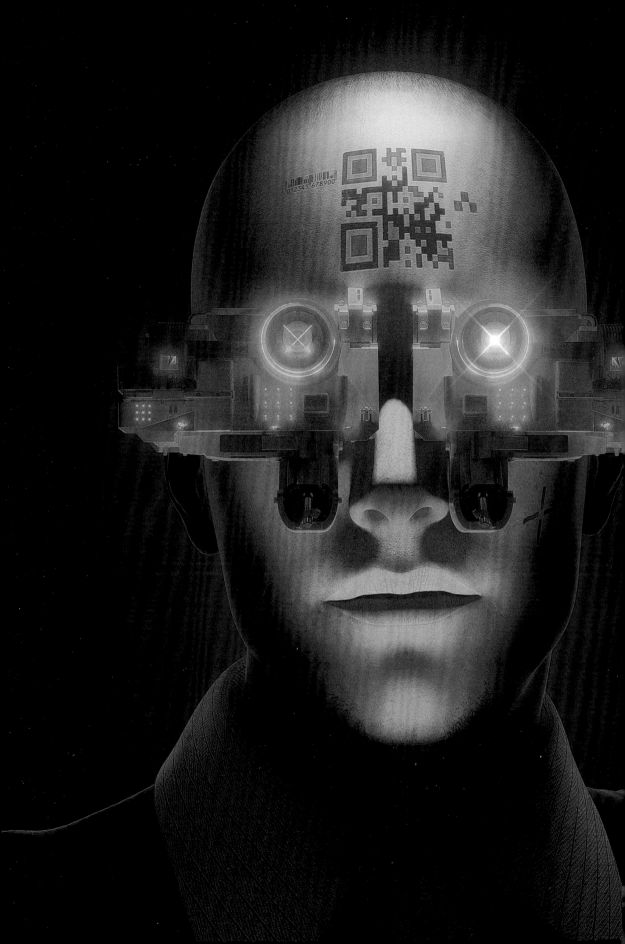

from the outside. As Joe Cannella, a former senior account manager at Google, once said, "You are given everything you could ever want, but it costs you the only things that actually matter in the end."[5] Although the mental and emotional chains dragging the modern tech workforce down remain invisible to people who don't understand the inner workings of the industry, employees from Silicon Valley's biggest tech corporations often verify this hidden secret—if you can get them to talk. There is no better proof of this than a statement released by former Amazon exec Charlie Kindel on April 30, 2018 about why he decided to leave Amazon, in which he stated: "The pace of the past five years has finally gotten to me and I simply need to catch my breath. I've recently been joking with folks that 'I used to get my adrenaline rush going heli-skiing. Now I just go into work.'"[6]

What's clear to see in Kindel's comments is that the toxic environments have been recreated—this time through mental entrapment. Of course, some are in a better position to walk away than others, but many have no choice other than to participate until their undoing. Even then their experiences are often kept quiet out of fear and embarrassment—fear of retaliation if they speak out, and embarrassment that if anyone finds out they may be looked down upon as weak or incapable of keeping up. Although workers often feel like it is their fault, what's really happening is that in pursuit of the perfect efficiency we have developed a world in which mathematical models created to drive profit are now micromanaging everything, including the labor market, to create model societies.

Hiring Employees

Finding good talent is one of the hardest things a business has to do. In fact, according to recent studies by the Center for American Progress, one of the biggest expenses for a company is employee turnover. For employees earning $75,000 a year or less, which represents 90 percent of the workforce, the cost of letting someone go and finding a replacement can be as high as 20 percent of that employee's salary.[7] Therefore, hiring the right people the first time is very important. To make the process easier, especially for large corporations, it makes sense to use a computer to help. Not only does it reduce the number of hours required by a human—time spent sifting through resumes, making calls, and setting up interviews—it also removes bias from the process, or at least that's what we would like to think. But if we take a deeper look we can see that's not necessarily the case.

A great example of what happens when algorithms control the hiring process is a story brought up by Cathy O'Neil in her book *Weapons of Math Destruction*, the case of Kyle Behm, a successful college student, looking for a part time job to help him make ends meet.[8] In 2012 Kyle decided to apply at Kroger, a corporate grocery chain, where one of his friends worked. For a high achiever like Kyle, the application process seemed to be a formality; but, despite having a connection and putting his best foot forward, he was denied the job. Although he could have gotten upset, he realized this was only one job. Maybe he just wasn't the right fit. Anyone could have been turned down. It could have gone to someone who applied before him. He resolved to carry forward without stressing about the lost opportunity.

The strange part to this story is not what happened with Kroger, but that the same thing happened to him after applying to Finish Line, Home Depot, Lowe's, PetSmart, Walgreens, and more. Even after applying to all of these places he was still unable to get a call back. It appeared as if none of these companies wanted to hire a well-to-do college student who had almost perfect SAT scores. But why? It turns out all of these companies were using an employee selection program developed by Kronos, a workforce management tool, to evaluate employees and eliminate people from the talent pool who didn't meet given requirements. Although use of such a selection program sounds logical, a closer look reveals something that might otherwise go unnoticed—human bias.

Consider how these systems work. If a business has been around for 20 years and wants to find only the best possible candidates, the ones who are likely to succeed in the company, it would make sense to feed it the résumés of the company's best employees. The machine could then do a statistical analysis of the qualities these people had in common and use the findings to sift through incoming applications. Although the system will undoubtedly discover qualified candidates, what goes unspoken is that these systems don't take into account the fact that humans hired every employee whose name was fed into the machine.

So while the resumes given to the machine may represent successful employees, the output of the algorithm is destined to reflect the same biases of the people who originally hired these employees. It's easy to see how this could lead to biases against women, people of color, graduates from less prestigious colleges, and so forth—despite the fact that selections made this way are written off as unbiased. And until this is recognized, the injustices created by these system will spread across any variable employers decide to target—even age, as dozens of companies, including Amazon, Verizon, UPS, and Facebook itself, are beginning to systematically eliminate older people, who they have deemed worthless in the modern economy, from the application pool.[9] Until we take a stand against this type of behavior, nothing will change. If anything, it will only get worse as these systems creep into other parts of the work.

Scheduling Staff

In the fall of 2016 I worked with a company intent on building an automated system to monitor store traffic and help create employee schedules based on customer traffic patterns. It seemed like a brilliant idea. If the people in charge knew that traffic would be low or high at certain hours of the day, the company could adjust the number of employees on staff accordingly, which saves the company money and ensures the employees don't get stuck with nothing to do. Initially it sounded like a win-win. But as I explored the idea more, I realized the potential for several negative effects that could easily appear if the system was given control without the proper oversight. The effects of this system, like all automated systems, are hard to see without tangible examples, but the story of Jannette Navarro, a college student struggling to pay her way through college and care for a child, illustrates just how ruthless these systems can be.[10]

As an employee at Starbucks, Jannette was subject to corporate schedule automation that had been put in place to automatically allocate hours in the most efficient way. But because the system lacked context and empathy, it often scheduled employees without regard to their lives outside of work. Thus Janette was regularly scheduled for what they called "clopenings"—a shift that required employees to stay late into the night in order to close the store, and then return early in the morning to open the next day. The system also tended to leave scheduling until the last minute, which put employees in a situation where they had little time to plan anything else—whether that be personal life, school work, or, for many low-wage workers and women like Janette, child care. To make matters worse these systems often keep employees under 30 hours a week so companies can avoid having to pay for health insurance and other benefits that full-time employees receive—a simple matter of business efficiency.

For a young mother like Janette trying to balance college, childcare, work, and more, this became unbearable. Janette was forced into a position where she was run ragged trying to keep up with the demands of the machine, putting her child in a position of incredible instability—an environment that has been scientifically proven to harm children. Thanks to undue circumstances created by a machine Janette was forced to drop out of school, which then made it harder to get a job long-term, ultimately subjecting her to a life of heightened anxiety and unwarranted insecurity. All of these factors combined create a vicious loop that impacts the lives of millions of low-wage workers and eliminates upward mobility, even when these people are trying to put their best foot forward.

Automated scheduling systems like the ones used by Starbucks and many others get sold to companies as a win-win solution to their scheduling woes. These systems are so favorable to business that companies like Uber, Lyft, TaskRabbit, and other sharing economy apps work entirely through on-demand, machine scheduling. And for those who don't have to suffer through the effects, the system is magically efficient. But for those at the wrath of the algorithm the story is not the same. For this reason, California recently passed the Fair Workweek Act, which requires businesses to give employees seven days' notice of their work schedule.[11] This law, which became effective in July 2018 and will be doubled in 2020, signifies the beginning of legislation necessary to keep these machines from running workers into the ground. Similar legislation has also been passed in Oregon, New York, and Seattle, but it is uncertain how long will it take for other states to follow suit.[12]

Evaluating Performance

The last part of the employment cycle that is getting automated is the evaluation process. Google is notorious for its analytical HR department, People Operations (POPS), which has done everything from evaluate the optimal amount of time for maternity leave after giving birth (five months), to how often employees should be reminded to contribute to their 401Ks, to the amount of time employees should spend in line at lunch.[13] Most of these calculations seem safe—even progressive. And, in fact, there are plenty of great examples in which data has pushed a company to improve issues that may otherwise be neglected. However, being too data-driven without being mindful of the outcomes can have its downfalls as well.

For example, what would your employer find out if they were monitoring your emails? Or your Slack messages? Or any data points that could be considered an invasion of your personal privacy without you knowing? Would you feel comfortable handing this data over? What if you didn't have a choice? Although employee evaluation systems are sold as analytical, unbiased, evaluation software, this what happens when it is installed. And this is when analytics go too far.

For instance, the main purpose of a system called Cataphora was to mine employee data—to read through corporate emails, monitor employee-to-employee interactions, and collect other intimate data to rate employees' value to the company. Similar programs have been created by Microsoft and IBM in order to streamline the HR department's work.[14][15] The danger in these systems isn't in the spirit of what they are trying to accomplish—a system truly capable of quantitatively analyzing the full spectrum of work behavior could do wonderful things for the world. The trouble, however, is that these systems are attempting to quantify parts of life that are not quantifiable. And then, despite the fact that their outputs are machine approximations, the results are used to determine the future of people's lives.

Despite how they're sold as end-to-end HR automation, they are incredibly limited in what they're capable of doing. They struggle to understand the nuances of life, such as what's a joke versus what's an idea. They struggle to identify the true source of valuable thoughts and ideas, because many of them originate during human-to-human interactions at lunch or at the coffee machine—not on the screen. And, ultimately, they struggle to understand many parts of life, which then get reduced to a fault upon logical analysis.

These system will never be able to evaluate the value of the person who is capable of making a tense situation calm by cracking a joke. They'll negate the value of the person who's willing to speak up against what everyone is doing when they see clear danger that no one else can. And they'll overlook the person who maintains the spirit of the company and catalyzes team chemistry. These are the things that take a team from good to great, and there's no quantitative assessment that will ever be able to evaluate these interactions appropriately. Nevertheless, companies are still deciding to give machines the power to determine who is valuable and who is not.

Human Elimination as a Service

Artificially-intelligent automated systems represent the business of our lifetime and lifetimes to come. Trying to avoid this fact would be like standing in the middle of the room and closing your eyes during a game of hide-and-seek, hoping nobody finds you. The tipping point has already been passed, and trying to stop the progress of these systems would be like trying to turn off the sun. But while such systems are destined to eliminate millions of jobs in the coming decades, we should not fear the people at the top of these corporations wanting to replicate human life. It's not of interest to them. There are plenty of people on the planet, most of which they don't like or get along with.

What the people at the top want are slaves—obedient workers that do what they're told, when they're told, without voicing complaints, taking days off, or receiving benefits—without any of the complications of managing irrational, emotional human beings. This isn't necessarily because they're evil, but because they want more control, in general. If the business leaders at these companies could guarantee that the company's stock price would go up every quarter, they would have more money in their pocket, and all of this would require less work, it would be a dream. And that's exactly what they're getting. Unfortunately, the process is forcing humans to comply—to work, act, and live—like machines, or face the risk of elimination. This has resulted in work conditions that are as mentally and emotionally abusive as the physically abusive conditions of the first Industrial Revolution. The difference this time is that if we can't keep up we'll be tossed aside as a tool that is no longer capable of doing its job because the machines are ready to take over.

CHAPTER 7

Automating Humanity

The last step in the transition is, of course, to replace human workers with robots—to automate humanity. But don't worry, the machines aren't coming to kill us; they're just coming to take our jobs and any stability that's left in our world. Oh, and by the way, we've been helping them do it the whole time, whether we realize it or not.

Tasks once done by humans are being automated at a rapid pace because the economic benefits of using robots and automated services are irresistible to corporations, investors, and anyone with a mind for doing more with less. Many people who remain unaware of what's happening tend to blame immigrants and outsourcing, but this simply is not the case. As proof, we can look to the fact that since 2009, worldwide sales of industrial robots have risen nearly 400 percent. And the five countries that represent more than 74 percent of this growth are China, Korea, Japan, the United States, and Germany.[1] These are countries that did commonly participate in outsourcing, but this dramatic rise in robotic workforces is due in large part to the fact that wages in the emerging markets where work was traditionally outsourced are now rising, due to demand. But instead of meeting the demands of these workers companies are deciding it's once again time to find cheaper solutions.[2] The impact of this behavior can clearly be seen if we look at the difference between the employment and corporate profits in the United States.

Looking at the trends from 1995 to 2011 as presented by Andrew McAfee of MIT, we can clearly see a strange development in the American economy. In the early 1990s, people were employed, and production matched. In the early 2000s the dot com bubble caused a crash in both the employment and the product, at least temporarily, but things quickly recovered. No long after, at about the 2003 mark, things start to get strange as corporate profits start soaring despite employment staying about the same. While part of this surely had to do with outsourcing, this also just so happens to be about the same time automated, artificially-intelligent systems started to get going.

Continuing on, we see the Great Recession where both the profits of these corporations and the rates of employment plummeted. The difference between the Great Recession and every other recession since World War II is that traditionally companies that laid off workers tended to rehire as soon as the economy recovered. But as the trend lines show, this time was different. While corporate profits recovered completely, and have since risen, employment remained low. How did this happen?

According to Robert Tercek, "One reason the US employment index responded so sluggishly after the 2009 crisis was that many of the workers who were laid off were not rehired. And they won't be. Ever. Those workers were replaced by machines." Some of this surely has to do with the outsourcing of jobs, but the greater truth is that work is increasingly getting handed off to robots. This lack of need for human workers is reinforced by an analysis from the Bureau of Labor Statistics (BLS), which announced that the private sector in the US did not regain jobs lost during the Great Recession until six years afterwards—a recovery effort that took longer than the previous two recessions combined.[3] In analyzing these trends what we're really seeing is that in the early to mid 2000s corporations discovered the opportunities automation would bring to business operations. But CEOs didn't want to just lay off hundreds of thousands of workers because it's hard to let that many people go at once, and even if that's not the issue, it's bad PR, so the

The Rebound that Stayed Flat

This chart represents a comparison of the United States GDP, employment–population ratio, and profits from 1995–2011. The analysis was performed by Andrew McAfee, principal research scientist at MIT, who studies how digital technologies are changing business, the economy, and society.[4]

■ All profits after tax
▧ Employment-population ratio

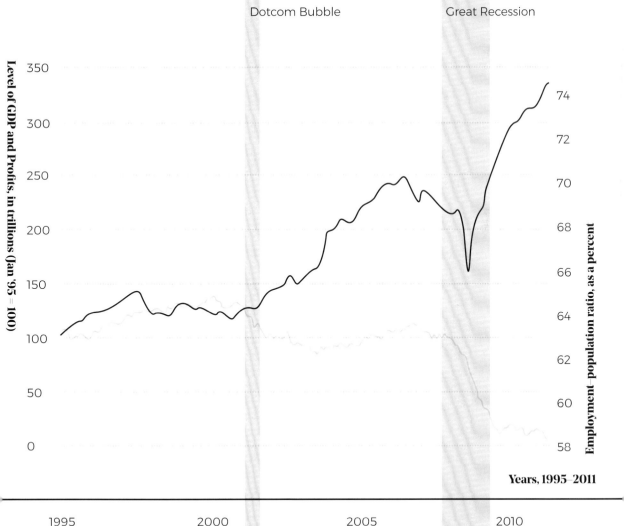

Dotcom Bubble   Great Recession

Level of GDP and Profits, in trillions (Jan '95 = 100)

350
300
250
200
150
100
50
0

Employment–population ratio, as a percent

74
72
70
68
66
64
62
60
58

Years, 1995–2011

1995   2000   2005   2010

people stayed. But what happened next is that the Great Recession gave these corporations a reason to lay off workers without having to take the blame. And what happened afterwards is pretty clear—the employees were never hired back. Although slight improvements have been made since McAfee released this discovery, things will never be the same.

Training Our Replacements

Now, and for the immediate future, human assistance and guidance will be necessary for machines to operate properly. Even the most powerful information bots such as Alexa, Siri, and Google Assistant need to be commanded at this point in time. But these hurdles will quickly be overcome. In the very near future bots will not only be able to respond to commands but will also be capable of generating ideas when given limitations, and eventually they'll do things on their own. As the cost of these systems continue to drop, the likelihood will only become greater that these machines replace people in the workplace. As Barack Obama stated in his farewell address in 2017, "The next wave of economic dislocations won't come from overseas, they will come from the relentless pace of automation that makes a lot of good, middle-class jobs obsolete."[5] Despite the fact that this might seem impossible to those who are unaware of what's happening, it is the truth. And we've been training these systems to replace us for decades. With each click, swipe, recorded conversation—whatever it is that we've been doing as we're connected to the internet—we've been feeding data to a system that will, more than likely, take our jobs someday. And though this fact may make us want to stop using the internet, we've already reached the point where there's no turning back.

A simple example of how we've been training machines we may not realize is ReCaptcha—you know, that thing at the end of all sign up forms that asks us to, "Type these two words," or, "Select all images with street signs," to prove, "I'm not a robot." What most people don't realize is that when we prove to the machine that we're human we're actually training machine intelligence for Google, which acquired ReCaptcha in 2009.[6] The words and phrases we receive are often from old books that Google's machine intelligence is trying to digitally document but can't understand because the words are smeared, faded from age, or otherwise difficult for a machine to identify. The photos are training a machine vision system to understand the world. With every response, Google can add it to their database and then compare the responses of billions of people to determine the appropriate response through statistical significance. This may sound like a system that's bound to be flawed, but with that many samples they can be completely confident about what should fill in the blanks.

ReCaptcha is a great example of how clever these companies are at getting people to train their system for them, but it doesn't seem like a threat to anyone's job, right? That's because it's not, really. But there are plenty of things happening on the inside of all these companies that you won't even know about until they're released to the public, and by then it's too late. For an example of how our jobs might be taken, let's examine how Google created Google Autodraw—a system built to turn anyone into an illustrator.

In order to collect training data for the system, Google created a fun game called Quick, Draw! in the summer of 2016 in which they asked people to, "Help teach [a neural network] by adding your drawings to the world's largest doodling data set." The point of the game was for users to sketch a given word in five seconds as best they could and then challenge friends to do better. But there's a limit to how good a drawing can be in five seconds, no matter how good of an artist someone is—and Google knew that. So what the game really did was give Google insights into what people generally think a word looks like while ensuring that nobody could make any world-changing works of art. Essentially, Google collected data that would teach a machine to recognize the worst sketches possible and that these horrible drawings represented the word it was attached to. Then, within nine months, Google took this data and reverse engineered it to create AutoDraw, which allows anyone to sketch what they want to illustrate and returns a wide array of professional-grade illustrations to use instead. For all intents and purposes this system eliminates the need for a professional illustrator if a company isn't willing or able to pay for one.

Till Bots Do Us Part

Essentially what's happening is that tech companies have turned billions of people into unpaid laborers who will then get cut out of the equation once the system is capable of doing their job. Anyone can argue about the value of humans, and that person wouldn't be entirely wrong, but the same argument could have been made for horses before Ford created the Model T in 1908. Despite the fact that horses were better at many things than cars—such as recognizing faces, jumping fences, and reciprocating love and affection—cars were powerfully more efficient and effective at the specific things humans needed them for and they were capable of performing these tasks in ways horses never could. Today horses are still around but their value is different, thanks to the mass production of automobiles. The same will be said of humans, thanks to artificial intelligence.

Like Google, companies across the world are jumping on this trend. Tesla is using human drivers to train their systems how to navigate the road; Adobe Sensei is using human designers connected to Adobe Creative Cloud to train automated creative systems; Stryker Medical, a company focused on medical automation, is slowly chipping away at the need for human doctors; and this is just a short list. In the next ten years we will see things that were previously only imaginable in sci-fi movies and TV shows, including robots that are intelligent enough to gain citizenship, as Hanson Robotics' Sophia did in Saudi Arabia in late 2017.[8] And in order to replace us, these machines won't need to be perfect, they'll just need to be better than us.

PART 2

How Did We Get Here?

We've been struck by a tidal wave of digital technologies that is ripping through our world like a chainsaw cutting through butter. Everything's happening so fast most people don't even know where it started or how it got this way to begin with. What's happening seems unreal. But it is real. And it's here to stay.

CHAPTER 8

Information, Technology, and Evolution

It's easy to forget where we came from, but if we allow ourselves to forget our origins, we're bound to repeat history. Although most people believe what's happening is abnormal and dangerous, it's completely logical considering history.

Communication, the ability to exchange information with meaning, has played a fundamental role in the creation of technology and in human evolution. Whether spoken out loud, written on paper, etched in stone, or digitally streamed through the internet, our ability to consume, use, and store information allows us to learn, transact, develop new tools, and, ultimately, it has shaped human history. In his book The Master Switch, Columbia University law professor and former senior advisor to the Federal Trade Commission Tim Wu makes the following statement about the impact communication technologies have had on society:

> **"** *When in the course of human affairs things go wrong, the root cause is often described as some failure to communicate, whether it be between husband and wife, a general and a front-line commander, a pilot and a radio controller, or among several nations. Better communications, it is believed, lead to better mutual understanding, perhaps a recognition of a shared humanity, and the avoidance of needless disaster. Perhaps it is for this reason that the advent of every new technology of communication always brings with it a hope for ameliorating all of the ills of society.*[1]

Examining the different ways humans have exchanged information over time can help us better understand why technologies have been produced. Through such an analysis, we can identify five main aspects of information exchange that have impacted evolution: context, speed of dissemination, reach, permanence, and organization.

Context is the sharing of language, life experiences, environment, or anything else that helps us understand the meaning of information being exchanged. For example, we might have inside jokes or other life experiences with friends that nobody else understands. Context gives us a better understanding of information by helping us understand where it came from.

Speed of dissemination refers to how quickly information moves from point A to point B. Consider that spoken words only travel as quickly as they can be shared between people, but tweets can spread across the globe with the push of a button. In this light, the faster information moves, the quicker it can be adopted by the masses.

Reach has to do with how far information spreads. For example, spoken words only travel as far as any given individual is capable of projecting their voice, but a blog post can reach millions of people around the globe. Similar to speed, the more people any piece of information reaches, the more opportunity it has to be adopted by the masses.

Permanence is how long-lasting information is. A stone tablet, for example, is much more permanent than spoken word, which only exists in our memory. The more permanent information is, the more opportunity it has to reach larger audiences, over time.

And lastly, Organization reflects how easy it is to find information. For example, it's easier to find a book in a library or on Google than it is to search for an elder tribesperson wandering the world nomadically. As information becomes more organized it is more likely to be found, which means less time working to discover it and more time using it.

When we analyze the way we've exchanged information over time with these factors in mind, it's easier to understand how we've evolved the way we have and why we're moving faster today than at any previous time in history.

Oral Cultures and Storytelling

The first form of information humans shared with was language, the roots of which were not spoken word or written scripts, but gestural movements, grunts, or other simple interactions that we today often fail to recognize as language. Despite the fact that such basic interactions may seem trivial compared to the ways we exchange information today, the impact language had on humanity when it was first created was profound. In its most basic form, language allowed people to take a thought (information in their head) and put it in someone else's brain—without Wi-Fi, wires, letters, diagrams, or any other tool.

Viewing language in this way, we can begin to understand why we commonly ask questions like, "Do you understand me?" and, "Do you get [or did you *receive*] what I'm saying?" Since each of us has a unique perception of the world, such phrases are how we check for mental alignment. What we're really asking the other person is, "Is the thing that was in my mind now in yours?" As Yuval Noah Harari says in *Sapiens: A Brief History of the World*: "The truly unique feature of our language is not its ability to transmit information about men and lions. Rather, it's the ability to transmit information about things that do not exist at all."[2] When two individuals are capable of coming to a mutual agreement about what something means, their worlds get a little closer to the same.

As amazing as this is, what often goes unconsidered is that spoken language was our main form of exchanging information for hundreds of years, which essentially meant life was one long game of telephone—and we all know how that turns out. To make the exchange of knowledge more permanent, people created memory devices called mnemonics; mental tools like rhymes, acronyms, and narrative storytelling. These tools, especially storytelling, laid the foundations of many societies. Without storytelling, people would not have common religions; there would be no debates about human rights issues; people would have no agreed value for currencies; they'd be unable to defend their beliefs; and large groups of strangers would be incapable of successfully cooperating in the pursuit of a common purpose.

Storytelling shapes our lives so much we get emotionally attached to the tales we tell ourselves. Think of how upset children get when they find out Santa Claus isn't real, or how emotional some adults get when someone suggests Superman could be a person of color. The words and images that create stories become so real they define our sense of reality. And once stories have been adopted as truth by enough people it becomes hard to change them, regardless of whether they are truth or fiction. Those who question the status quo often become regarded as crazy, mentally ill or, worse yet, conspiracy theorists—out of touch with what we call "reality" despite the fact that our reality is an alternative story in its own right. People react this way because such challenges can destroy the narrative that gives our world consistency and stability.

Perhaps the greatest strength of oral communication has been that there is much context within the information being shared. Two individuals speaking to each other face-to-face often shared a history and language, and were also capable of receiving other cues based on body language, all of which made communication easier. The weaknesses of oral communication are that information is only as permanent and organized as our memory and can only be shared as far as we're capable of yelling, limiting our understanding of the world.

Seeing Sounds

After millions of years, societies began to develop written languages. While cave paintings have been traced back as far as 20000 BCE, meaningful communication through systems of symbols took quite some time to develop. Estimates for the beginning of written language range anywhere from 3500 to 3000 BC and are typically attributed to the Sumerians. However, it wasn't until sometime between 3000 to 2500 BC that language evolved into what the Sumerians called cuneiform, a more advanced written language similar to what we're used to today. Around the same time the Sumerians developed cuneiform, the Egyptians developed hieroglyphics. Then, about 1,500 years later, around 1200 BC, China developed its own language, followed shortly after by civilizations in Central America that developed their own languages, around 1000–500 BC.[3]

What's incredible about written language is that each letter, word, and sentence we read is a visual representation of the sounds or body movements we would otherwise have to make to communicate. If individuals are not trained in interpreting these symbols, participating in society becomes much more difficult. The progress our world has made due to what we call "literacy" is astonishing.

Having the ability to communicate with letters, words, phrases, and other visual symbols meant that people were now capable of storing thoughts—invisible, otherwise meaningless things—in a physical dimension that could be referenced years later. It also allowed for increased privacy to be had between conversational participants because they could communicate without making noises. Because of this, literacy remained the root of wealth and power for thousands of years as only the wealthy people could afford to learn how to read and write, enabling them to participate in activities others could not. Thus, knowing how to communicate through written script was powerful, often greater than money itself. The most dominant organization to control the people through disparities in literacy was the Church, which for centuries instructed people how they should live their lives before they were literate enough to form their own opinions.

Beyond this power struggle, the ability to read and write permitted an incredible increase in permanence of information. It allowed people to learn from previous generations, long after they had passed.

It allowed communities to record ongoings in order to improve collective understanding. And it altered the power structures of the world. However, it wasn't until the invention of the printing press that people began to see revolutionary changes in the distribution and storage of information.

Print and Repeat

About 5,000 years after the invention of written language, in 1455, a man named Johahnes Gutenberg invented a movable-type printing press that would revolutionize the way people used and stored information by allowing it to be replicated much faster, which meant it could be sold at a fraction of the price.[4] This gave more people access to printed information, which resulted in great increases in literacy. It also created a demand for places to store all this information, which resulted in more libraries. Although libraries had existed before the printing press, they now had value to the general public—as a publicly accessible information center, not so different than what Google is to us today. The difference is that no single library had nearly as much information as Google does, and, in order to find information, people had to physically search the library, not type into an input field on a web page.

As literacy rates accelerated and more people consumed more information, societies began to form around information. For example, the Bible allowed people to interpret the word of God in their own home, which took the power away from the Church and allowed people to create their own societies based around different interpretations of the Bible. While this caused dramatic changes to society, the desire to connect to others has been around since the dawn of time. It is the same thinking that leads us to participate in online forums, Facebook groups, and other digital platforms to stay connected with people. However, back then there weren't easy ways to connect and stay connected, which created a need that resulted in services like the United States Postal Service, Royal Mail, and Pony Express, among others.

Although information was still confined to geographic, linguistic, and literacy barriers, the printing press changed the way information was consumed, used, and stored across the world. The downfall was that with increased reach and speed of dissemination, communication began to lose a critical variable—context. This is crucial because if the context of information is misunderstood it becomes much more difficult to understand meaning, and if the meaning can not be understood, then the information is meaningless.

Can You Hear Me Now?

Several hundred years after the invention of the Gutenberg press, researchers began studying ways to communicate faster, without the need to print information on paper, package it, and send it through the mail. Their search led them to explore how information could be transmitted through what are called Hertzian waves and, just before the beginning of the 20th century, they made a breakthrough with the invention of the radio, which made its first transatlantic broadcast in 1901.[5] This was a major turning point in the history of communication technologies, but it wasn't until several decades later that radio would make its public debut on July 2, 1921, as thousands of people stormed their local radio halls to listen to "The Fight of the Century" between Jack Dempsey and Georges Carpentier. This event would not only result in a historic fight, but would also mark the first time there were more people experiencing an event remotely than in person—a concept we take for granted today because it has become so familiar.[6] Before radio though, it wasn't possible to broadcast something in New York and be heard by a large audience in Chicago, or anywhere else with a receiver, within such a short amount of time. However, while the speed, reach, permanence, and organization of information continued to increase during the age of radio, context continued to decrease as audiences grew larger than ever before and visuals disappeared.

The Sight Heard Round the World

Not long after the invention of radio, society underwent another revolution in communication, but this time visuals accompanied the sounds. In 1927 Philo Taylor Farnsworth invented a way to stitch sounds and pictures together, and called it television.[7] With this new technology, people were capable of seeing things around the world as if they were present, and to experience history as if they had been there when it happened. This new way of communicating had an immediate influence on society, and by 1960, 88 percent of American households had a television set; only a small fraction still relied on radio as their main channel of communication.

The impact of this new technology was demonstrated on September 26, 1960, when John F. Kennedy and Richard Nixon stepped on stage for the first ever presidential debate to be broadcast on national television. That night, an estimated 74 million people tuned in to watch the now-historic debate.[8] When all was said and done, the power of combining audible and visual stimuli became clear. Those who listened on radio—the people who could hear but not see the debate—thought Nixon won, but those who watched on TV—the people who could both see and hear the debate—thought Kennedy won. As history is told it is believed that Kennedy won the election that night because of how young, healthy, and handsome he appeared next to Nixon—things that never would have been scrutinized through audio alone.

The emergence of audio/video (A/V) technologies like TV enabled audiences to have a fuller understanding of the world by engaging additional senses. This gave audiences an increased level of context they were incapable of experiencing before. It also allowed these context-rich messages to be sent across the globe at the speed of light, to millions of people at once, while simultaneously recording them as history for the archives. However, just as with radio, context was also lost as broadcasts reached larger audiences who may or may not have understood what they were seeing.

The Wild Wild Web

The next major revolution in communication came in 1991 when British physicist Tim Berners-Lee launched the first website on the "Internet." Not long after, the internet became the farthest reaching, most neurotically organized communication tool the world has ever seen. Six years later, in 1997, there were over 1 million websites on the internet. Today, less than 30 years since the first website, there are more than 1 billion websites.[9] What's more, through these connections people could now communicate from anywhere, to anywhere, through both sight and sound, without an intermediary.

As revolutionary as the internet was in its early days, it has only evolved and gained even more importance over time. Although more than half the world still doesn't have access to the internet, it is still the largest and fastest-growing communication technology in the history of the world—a speed that is only getting faster as it evolves.[10] Consider, for example, that in the 1990s people still had to wait for the dial-up modem, plug their camera into a computer to upload photos, and exchange information on floppy disks or other external memory devices. Now, with wireless internet, camera phones, and cloud storage, this all happens in real-time. Moreover, although most people don't think about it this way, we are also actively recording history every time we do anything on the internet. Our timeline may only represent a history of our life, but all of our timelines combined represent a history of the world—at least a partial one. Through our everyday actions we have collectively created the largest centralized source of immediately searchable information in the history of the world, and as long as there are servers plugged in with space to store our information, that database of knowledge will only continue to grow.

With no geographic, time, or language barriers (thanks to translation services) confining our ability to share information, anyone using the internet can now access any piece of information from any point in recorded history as long as they have the appropriate link. This has brought us to a point at which our communication difficulties no longer reside in the act of communication itself, but have instead become an issue of expressing meaning across barriers of context. As a reaction to this, individual languages are beginning to break down in favor of a more global language based in symbols and emojis, compound words and acronyms, as well as GIFs, memes, and other pieces of interactive media. In short, we are returning to the origins of communication to overcome the hurdles of modern communication.

Our new, abbreviated language is also being complemented by machines that can communicate for us. Whether this takes the form of machine-to-machine (M2M) communication, in which machines communicate with each other through Application Program Interfaces (APIs), or machine-to-human communication (M2H), as we see with chatbots and other automated services, the result is that society is reaching a point at which we won't even think about information anymore, which is why we've to begun to move so fast. This is why our ability to consume, use, and store information is faster than any point in history, which is clearly reflected in the world's production value of over $70 trillion in 2017—a value that is 280 times what it was 500 years ago, despite the fact that the global population has only increased by a factor of fourteen.[11][12][13] As Yuval Noah Harari states in his book *Sapiens*:

> " *Centuries ago human knowledge increased slowly, so politics and economics changed at a leisurely pace too. Today our knowledge is increasing at breakneck speed, and theoretically we should understand the world better and better. But the very opposite is happening. Our newfound knowledge leads to faster economic, social and political changes; in an attempt to understand what is happening, we accelerate the accumulation of knowledge, which leads only to faster and greater upheavals. Consequently we are less and less able to make sense of the present or forecast the future.*[14]

While the future is impossible to predict with perfect precision, what we can be sure of is that more change will come. We are not a society that remains satisfied for long, and with any insight into what's been happening over the past 10 years or so, it is easy to see that we are on the cusp of another breakthrough. Where we're headed is a world in which more of our senses will become engaged—a world where we can not only see and hear information, but fully immerse ourselves in it, allowing us to communicate without saying a word.

CHAPTER 9

Billions of Brains as One

Artificial intelligence is causing a second Industrial Revolution in front of our eyes. However, instead of catalyzing human muscle as the previous revolution did, we're catalyzing human intelligence. While it is incredible to watch, these systems still have their flaws just like the humans that created them. It's important we recognize these flaws before we give the machines our full-hearted trust.

Fundamentally, intelligence is the ability to consume, store, and use information. As humans, we're relatively good at consuming information because our sensing organs—eyes, ears, nose, tongue, and skin—have been evolving for millions of years. However, when it comes to using information we don't always perform so impressively because the way information is interpreted is subjective, based on life experiences and knowledge, which means some people are better at interpreting certain types of information than others. And when it comes to storing information, that's where human intelligence really drops off. This is completely the opposite for computers, which have always been highly effective at storing information but not so good at consuming or using it. Until recently, computers have had to rely on humans to input information and write programs telling them how to use it.

Today, that story is different thanks to a wide array of sensors embedded in our machines that allow them to consume and use information in a manner similar to us, sometimes better. Cameras allow machines to see. Microphones allow them to hear. Accelerometers (sensors that monitor device acceleration in different directions) and haptic sensors allow them to sense directional equilibrium, multidirectional speed, and various forms of force and touch, among many other things. Not only are these sensors so tiny that we barely notice them, they're also so powerful they can consume information at a rate beyond any human ability, with much greater accuracy. And, thanks to advancements in computer science, the programs and algorithms being created today have surpassed our abilities to use information in many ways. Soon, more human senses will be replicated and machines will have the full capacity to consume information like people, but more efficiently. This relatively recent ability of machines to perceive the world around them is what has led us into an era of advanced artificial intelligence. The last hurdle that needs to be overcome will be in defining how machines use information, but that day will come.

Although artificial Intelligence (AI) has only reached the mainstream consciousness in the past several years, the concept of extending human intelligence into tools is something that has been around for centuries. Think about how the abacus took math out of our heads and into physical form, later allowing us to create modern calculators and computers. Think of how books used to be made by laying out and editing text by hand until digital word processors and design software came along. Think of the manual transmission versus the automatic transmission in cars. Even a computer, which we all now know as a thinking machine, was at one point in time a human job title before we had the machines we do today. These are all simple forms of automation that have redefined labor markets, but they're not so different than what's being done today with AI.

Despite how complex people tend to make it sound, artificial intelligence is exactly what it sounds like—an artificial replication of some form of intelligence. Sure, the math behind it might be more complex than in previous generations, and the machines themselves may be more powerful, but no matter how complex anyone may want to make it sound, artificially intelligent machines all have a program stored internally that allows them to consume information (in the form of settings or commands) and use that information to accomplish specific tasks, thereby automating processes that traditionally required human intelligence or participation, or both. Each system is simply taking the process out of our organic brains and automating that process in an inorganic object that requires as little human intervention as possible—sometimes completely eliminating the need for humans to acquire those skills.

Some of the simplest forms of artificially intelligent machines that we still use today include the thermostat, which uses sensors to monitor the temperature of a given environment (consume) and changes the temperature of a given environment (use) based on changes in temperature relative to a setting specified by a human (store); the washing machine, which washes and dries our clothes (use) once we tell it what to do (consume) based on its preprogrammed capabilities (store); and the ATM, which automates the process of banking transactions by taking commands from a human (consume) to perform actions that traditionally would require a banker (use) and track the interaction for comparison with future interactions, whether with the ATM, a banker, or any other bank interface (store). When we remove the technical jargon associated with the modern perception of artificial intelligence, the concept doesn't seem so complex, does it?

In modern terms, basic machines like these could be considered specialized forms of artificial intelligence—systems that only do one thing but do it very well every time. However, ask these machines to do anything outside of their preprogrammed capabilities and the intelligence fails to exist, just like specialized forms of artificial intelligence being created in Silicon Valley today. We may not perceive many of the simpler, basic forms of artificial intelligence the same way we perceive the machines being made today, but that's only because our perception of impressiveness is relative to the timeline that is human history. In 20 years we'll feel the same way about the machines being created today because AI is nowhere near its potential, which, at its most impressive form, will be like billions of brains replicated as one in every device connected to the internet across the world—on demand.

The Irony of AI

Despite the incredible potential of AI, there is also one fundamental flaw that tends to go overlooked. When it comes to learning, we always learn from somebody, whether that is our parents, a teacher, a coach, or a pack of wild animals. Every experience is an entry in the knowledge database that is our life. For a machine, its knowledge exists in the code base that structures its programs and the data it is given to understand the world. This, however, is also where its flaws exist. Although it often goes unrecognized, these two assets—code and data—are not naturally occurring elements. Unlike ironc, zinc, or magnesium, these assets always come from somewhere, and that somewhere is human. This means that, just like a child, the things a machine learns along the way may be biased or flawed. The key difference though, is the speed at which each learns, which is where things can become dangerous.

Humans gain experience by practicing over time, and although they may hire a coach or a mentor that accelerates the learning curve it takes quite a bit of time for most humans to become an expert at anything. By contrast, a machine learns much faster, sometimes almost instantly. Install an application and the computer automatically knows what to do the second it's opened. Add a plugin to the browser and its operations are instantly upgraded. Feed data to a machine learning algorithm and it might take some time to process, but relative to the human learning curve it's much faster. This is great for industrial efficiency, but with greater speed comes less room for error, which is where the danger resides. This is a detail that all too often gets overlooked in favor of achieving higher performance metrics.

Despite what any company may try to claim about the safety or unbiased, data-driven perception of their machines, we must remember that intelligence always comes from somewhere. Just as there has never been a child born who immediately knew how to live independently, there has never been a machine that was plugged in and understood how to operate. The greatest threat of artificial intelligence is not its ability to take over the world, but that we treat such systems as fully independent, all-knowing, and unbiased even though they function according to instructions defined by humans and data created by humans. As much as Silicon Valley may want to tell us otherwise, there are many parts of life that will never be able to be quantified and replicated into a binary program—at least not yet.

CHAPTER 10

Can't Stop, Won't Stop

This new Industrial Revolution is here to stay. It's not going to slow down anytime soon. In fact, if we analyze the industrial growth of digital products in the past few decades, what's clear to see is that growth on the internet is similar to a virus. There are several reasons why this is happening, but the big picture is not easy to understand unless you know what you're looking for. However, once you see it, it's hard to forget.

Computer scientists have been pursuing artificial intelligence since before the 1950s, but it wasn't until a couple years into the 50s that AI
started to show promise. In 1952 a computer was built that could play
tic-tac-toe.[1] In 1964 Joseph Weizenbaum created ELIZA, the first artificially intelligent psychiatrist.[2] By 1996 Microsoft launched Clippy, the
first intelligent office assistant.[3] These systems were nothing spectacular compared to today's technologies, but to the people of previous
generations, they were impressive. However, the progress that has been
made since is incredible.

In 1997, some 45 years after the beginning of artificial intelligence programming, a machine named Deep Blue, built by IBM, beat
then World Chess Champion Garry Kasparov at chess, a game that requires a high level of intelligence and sequenced logic.[4] Fourteen years
later, in 2011, IBM's Watson beat Jeopardy champions.[5] At this point
artificial intelligence was operating beyond patterned logic and had
moved into the realm of human understanding. Four years later, in
2015, DeepMind's AlphaGo beat champions of Go, the most difficult
game in the world, despite the fact that experts had predicted this
would take a decade to achieve only a year earlier, in 2014.[6][7] With this
achievement, machines made the leap beyond human understanding
to intuition.

These examples demonstrate just how quickly technology is
evolving. In less than a lifetime artificial intelligence has gone from being able to compete in a child's game to conquering the pinnacle of
strategic thought. To understand why this is happening so fast and why
it's not going to slow down anytime soon, three key factors must be
considered: processing power and storage capacity, mass data collection, and algorithmic efficiency.

Processing Power and Storage Capacity

1 You don't need to
know what a teraflop
is, just compare
the rate of change
between the num-
bers you see is this
chapter.

In 1996, the US government's Accelerated Strategic Computing Initiative (ASCI) Red was the world's fastest computer. It was the first
computer capable of performing one trillion floating point operations
(1 teraflop) per second, but cost $55 million to develop and required
nearly 1,600 square feet to be laid out and assembled. By 1997 it nearly
doubled its performance capabilities, processing up to 1.8 teraflops per
second.[8] This amount of power in one machine was unprecedented in
computing history. However, not long after the ASCI Red had reached
1.8 teraflops another machine did as well—the Sony Playstation 3.

Although the computing power was nothing more than the
ASCI Red, it cost less—$500 instead of $55 million—and was smaller,
about the size of a couch pillow rather than the size of a three-bedroom
apartment.[9] Amazingly, this progress occurred in just under a decade,
and has only accelerated over time due to a concept called Moore's
Law, which Gordon Moore, the co-founder of Intel, brought forth in a
1965 magazine article titled, "Cramming More Components onto Integrated Circuits." In it, Moore wrote about how the amount of computing power people could buy for one dollar was doubling each year, and
speculated that this could very well continue for the next decade.[10] As

it turns out, the doubling predicted by Moore is still happening today, more than five decades later,

In this time period we've gone from chips with less than one thousand transistors, to chips with more than 30 billion.[11] A $700 smartphone is now more powerful than the world's fastest supercomputer from the early 1990s.[12] And in terms of teraflop performance, Google now has Tensor Processing Units (TPUs) performing more than one hundred times faster than what was possible just two decades ago—a feat they've stated "may look small if the promise of quantum computing comes to fruition."[13] The speed at which technologies are evolving has become so fast, many people are unable to logically comprehend it anymore. This growth in processing power and storage capacity coupled with the amount of data being processed each day, has been critical to the acceleration of artificial intelligence.

Mass Data Collection

Since 2005, EMC, a data storage company, has sponsored the IDC Digital Universe Study to examine the amount of information generated across all digital networks in the world. In 2005, they discovered more than 130 exabytes of data were being created each year.[15] Since then that number has doubled every 18 months, thanks in large part to companies trying (and succeeding) to mine as much data as possible. For example, Twitter reported over 350,000 tweets per minute in 2013.[16] YouTube reported more than 400 hours of video being uploaded to the platform every minute in 2015.[17] And in early 2018 Facebook reported that there are 510,000 comments posted, 293,000 statuses updated, and 136,000 photos uploaded every 60 seconds.[18] To put this in perspective, we are now creating more information in a day or two than our ancestors created from the dawn of human history until 2003.[19] At this rate, EMC predicts that by 2020 there will be more than 44,000 exabytes of data in the digital universe, an amount that Robert Tercek did a great job of expressing in his book *Vaporized: Solid Strategies or Success in a Dematerialized World.*[20]

In order to visualize this he asks readers to imagine a 1 GB flash drive. To represent the 44,000 EB of data we are expected to create by 2020, we would need to fill 4.4 billion bedrooms from floor to ceiling with 1 GB flash drives. Even this is unimaginable, but with this it's easier to understand why we operate in cloud platforms—because things are moving so fast that upgrading hardware in order to keep up with storage needs isn't efficient from a monetary perspective, and creating a single building that could hold all of this information as it grows and expands would literally be impossible.[21]

Algorithmic Efficiency

Increases in processing power and storage capacity have allowed machines to store all of the data we're creating, and increases in data collection have given machines the knowledge to operate, but Ray Kurzweil's "Law of Accelerating Returns," the idea that as things become

2 The average price of a computer fell 95 percent from 1998 to 2013.[14]

3 10,000,000,000,00 0,000,000,000,000, 000,000,000,000,0 00,000,000,000,00 0,000,000,000,000, 000,000,000,000,0 00,000,000,000,00 0,000,000,000X

This number, above, is the performance increase Sergey Brin expects if Alphabet can engineer a 333 qubit error-corrected quantum computer.

more organized and technologically advanced, we're capable of doing more in less time is what will continue to keep us going beyond any physical limitations.[22] This is because the concept has less to do with physical infrastructure or the amount of data being collected and more to do with the organization of life in general.

For example, Google Search allows us to find and digest information faster than looking in a library, which is still faster than finding the local shaman in the next village. Planes allow us to move faster than cars, which permit us to move faster than on foot. Instant messaging services allow us to communicate faster than by mail, which lets us communicate faster than in person. With these advances in both technology and organization, over time, we're not only seeing paradigm shifts occur more often, but greater leaps of technological progress within these shorter time periods as well. This synergistic effect is what Kurzweil calls "accelerated returns," and because of this, he predicts that people in the 21st century won't experience one hundred years of progress like the linear model of returns predicts; instead, we will see more along the lines of 20,000 years of progress—an exceedingly accelerated rate of return.

Will It Slow Down?

Exponential growth is something most people have trouble comprehending because it can quickly result in numbers so large that we're incapable of understanding them. To help people understand, Ray Kurzweil tells a story of two men—the inventor of chess and his patron, the emperor of China who, in an attempt to reward the inventor for the genius, invites the inventor to name his price. To the emperor's surprise, the inventor asked only for rice to feed his family and that the emperor use the chessboard to measure the quantity, with each step across the board doubling the amount of rice from the previous step.

" *As the emperor and the inventor went through the first half of the chessboard, things were fairly uneventful. The inventor was given spoonfuls of rice, then bowls of rice, then barrels. By the end of the first half of the chessboard, the inventor had accumulated one large field's worth (4 billion grains), and the emperor did start to take notice. It was as they progressed through the second half of the chessboard that the situation quickly deteriorated.*[23]

In one version of the story, the emperor goes bankrupt because the 63 doublings represented on the board totaled 18 million trillion grains of rice. In another version, the inventor loses his head; either way, this example demonstrates just how quickly exponential growth moves. Today, the rate at which technology is progressing is akin to the second half of the chessboard in Kurzweil's story, and there's no slowing down. Even if we hit the upper limits of processing power and storage capacity; even if we can't collect anymore data; even if we get stuck on the best way to organize information, it will continue to grow at an incredible pace because there are trillions of dollars to be made by connecting everything to the internet.

In fact, it is estimated that more than 75 billion devices will be connected to the internet by 2025, up from 15 billion in 2015.[24] In tandem with the growth of connected devices, the IDC has predicted that by 2020 we will be creating 44 zettabytes (ZB) of information each year—10 times what was created in 2013.[25] And with all of this extra data it's estimated that the value of the industry will skyrocket to more than $14.4 trillion by 2022, an increase from the $486 billion value of the market in 2013.[26] [27] Data mining is a huge business and connecting all of these devices will only make it easier to do. Unless the marketplace implodes, there's no logical reason to assume such developments will slow down anytime soon, meaning we are likely to achieve what Erik Brynjolfsson and Andrew McAfee describe in The Second Machine Age: "Our generation will likely have the good fortune to experience two of the most amazing events in history: the creation of true machine intelligence and the connection of all humans via a common digital network, transforming the planet's economies."[28] And if this is done safely, these technologies will allow the world to flourish.

PART 3

What Could the Future Be?

We're now capable of doing things previously only possible in sci-fi films. Although there's plenty to be concerned about with the current state of technology, the overwhelming majority of what's happened has been proven to be good for society. And the good news is that if we do this well, there is an abundance of benefits still to come.

CHAPTER 11

Amplified Intelligence

The mechanical tools we created with steam and electricity allowed us to amplify our natural physical abilities—we could move faster, travel longer distances, lift heavier weights, etc. The digital tools we're creating today will enable us to amplify our natural mental abilities the same way, and if it's done well, it will be incredible.

The invention of the steam engine allowed humans to perform physical work at a pace impossible for previous generations—the same way artificially intelligent machines will allow us to accelerate our cognitive abilities. The idea of amplified human intelligence is something Joseph Carl Robnett (J. C. R.) Licklider contemplated in his paper "Man-Computer Symbiosis" in 1950. [1] Not long after, Doug Engelbart's "the mother of all demos" gave the world a preview of how this might happen, in which, Engelbart used a mouse to interact with another individual on a hypertextual interface from separate locations. [2] This doesn't sound like anything spectacular given the technologies we have today, but consider this was done in 1968—16 years before the original Macintosh, 35 years before Skype, and 44 years before Google Docs. That same year, Licklider and fellow scientist Robert Taylor made the following prediction in their paper titled "The Computer as a Communication Device":

> *We believe that we are entering a technological age in which we will be able to interact with the richness of living information—not merely the passive way that we have become accustomed to using books and libraries, but as active participants in an ongoing process, bringing something to it through our interactions with it, and not simply receiving something from it by our connection to it.* [3]

Although we've been able to create interactive informational experiences on the web for years, the convergence of web technologies, large amounts of data, and virtual reality tools has brought this experience to life in a way previous generations of computer scientists could only dream about. We can now, as Licklider and Taylor predicted, "interact with the richness of living information." Today, we are capable of creating tools that accurately reflect many of our various senses, which enable fully immersive experiences that make information not only more enjoyable to digest but also much easier. Instead of having to learn through either an audio, visual, or kinesthetic experience, we can now engage all three at once. Through these experiences we will be capable of amplifying human intelligence beyond anything the world has seen before.

Improved Understanding of the World

One of the simplest ways to comprehend just how significantly the internet has impacted our world is to review the evolution of maps. For centuries people relied on cartographers to explore the world and translate their findings into maps that could be used to navigate. As technologies evolved, maps could be created with greater accuracy thanks to satellite imagery. But now, as more devices are connected to the internet, understanding the context of the world has never been easier, because of an idea called crowdsourced knowledge, in which every person connected to the system creates data points that inform a central database. Through this we can create real-time, interactive data experiences like Google Maps.

From real-time traffic patterns on the road, to real-time data about how busy a store is, to recommendations about what's nearby, Google Maps takes information from billions of people connected to the internet and allows us to understand the world from a God-like perspective. Every time we take a picture of something and share it on the internet, log our location to social media, or use a mapping service to get us from point A to point B, we help systems like Google Maps understand the world. Like a canyon carved by powerful waters full of little grains of sand, the compounding effects of billions of people connected to the internet has resulted in an understanding of the world that is so high-fidelity we don't even realize it. Although this gets taken for granted, it is clear to see as we zoom in on Maps and watch as it transitions from a traditional, flat map to a 3D rendering of Earth.

Think about how crazy that is. Do you remember when Google Earth came out in 2001 and your desktop computer struggled to render a low-fidelity, static version of the planet? Now, seventeen years later, each of us have a high-fidelity 3D rendering accompanied by real-time data that runs on our cell phones! Even more incredible, perhaps, is the fact that anyone who owns a Google Pixel 2 can make a real-time 3D rendering of the earth—with cloud patterns, lightning strikes, real-time sunrise and sunset, and more—their phone's background wallpaper, without impacting the phone's performance.

Soon, with what's coming to Google Maps, people will be able to experience an augmented version of earth that allows them to simply hold their phone up and navigate the world like a video game.[4] But Google isn't alone. Microsoft is doing similar things with AI, using it to help document and recreate historical sites around the world.[5] Many new cars now come with a heads up display (HUD) that allows drivers to navigate the world the same way fighter pilots do.[6][7][8] And, as of 2015, a company has even created a system called Be My Eyes, which allows millions of people around the world to act as on-demand eyeballs for those who can't see.[9] Never in the history of humankind have we been capable of understanding systems this large in real-time, like we can today. Over time, these systems will enable a world of amplified human intelligence like no other time in human history.

Improved Health Care

Another place artificial intelligence has incredible potential to help humanity is within health care. One of the biggest differences between mediocre and world-class doctors is their ability to correctly diagnose health problems, which, considering the complexity of life, can be incredibly challenging. According to research, it would take more than 160 hours each week for doctors to consume all the new medical information published weekly.[10] Then, they'd have to understand how the information fits into what they already knew and apply it without forgetting a detail. Although impossible now, this would be possible if assisted by artificially intelligent machines.

Imagine if, instead of taking days, weeks, or months to come up with the right diagnosis, a machine could look through all the diagnoses ever made, match a patient's symptoms with the medical history of the world, and give the doctor a short list of all possible diagnoses, ranked from most to least likely. All any doctor would have to do is enter all the symptoms, let the machine do its job, then use their professional expertise to select the diagnoses from the list of options and test patients in much more efficient manners. This may sound unreal, but it's exactly what's happening in hospitals around the globe thanks to medical assistants like IBM Watson, Google Brain, and other AI assistants, which have already been proven to be better at recognizing pneumonia, lung and breast cancer, and heart disease, among many other conditions.[11][12][13][14] Together, along with human experts, these systems will enhance doctors' ability to make the correct diagnosis in a shorter time frame than ever before.

Beyond diagnosing, AI systems can also help doctors be more aware of each patient's entire medical history, both personal and familial; how their history relates to everyone else in the world; and, as more genomic data becomes accessible, how this all intertwines with details of the human genome. With this information, machines will be able to do things no human ever could, in a fraction of the time. Pair this with connected health devices like GlowCap and MedReady that extend the doctors' ability to make sure patients are keeping up with their orders, and the level of medical care will far exceed any level care being offered in today's medical service industry.[15][16]

Improved Public Safety

Despite their ability to improve our lives, many digital tools have become so deeply embedded in our lives that we forget how much better off we are because of them. Just having a cell phone creates a level of personal public safety beyond that of previous generations. Combined with the internet, an incredible amount of public good can, and is being, done by these systems. Facebook, for example, allows people to check in during emergency situations to let their friends and family know they're safe. Google's Project Jigsaw leverages Google's world-class machine learning capabilities to "tackle some of the toughest global security challenges facing the world today"—everything from mitigating the threats of digital attacks and countering violent extremism to protecting people from online harassment.[17] In addition to these examples from tech giants, various projects like Revolar, Companion, and Watch Over Me are being created by startup companies around the world in an attempt to make our world safer.[18][19][20] As the capabilities of these products and systems become more refined over time, the level of public safety will only rise exponentially as long as the work is done with ethics in mind.

Thinking with Machines

Steve Jobs once called the computer a bicycle for the mind—a tool that allows us to operate beyond our natural human capabilities, but still requires work by the human using it.[21] Similar sentiments have more recently been expressed by many, including Robert Tercek, who has stated, "If we think of mechanical robots as non-biological 'brawn,' then artificial intelligence can be likened to a non-biological brain."[22] Moving forward it will be more important than ever to recognize the value humans bring to this partnership. What machines are good at is analyzing massive amounts of information in a logical manner. What humans are good at is interpreting nuances. The key to creating a meaningful future will be that, as Alibaba Founder Jack Ma has said, "machines should only do what humans cannot."[23] The tools created in coming years should complement and amplify human intelligence, not replace it. Together, humans and machines can work in harmony, and through this symbiotic relationship we can create a world that previously has only been possible in our imaginations.

CHAPTER 12

Accessible and Inclusive Technologies

Our natural world is filled with an incredible array of biologically created technologies. From opposable thumbs that allow us to grasp things, to wings that allow birds to easily maneuver through the air, to gills that allow fish to breathe underwater, evolution has been significantly beneficial to all life on this planet. As we move into the future, these organic technologies created by biological evolution will be recreated in various ways that enable us to do things we never thought were possible.

Enhancing humanity's biological capabilities with machines sounds like a future of human cyborgs to many people, but a brief reflection on human history clearly demonstrates this transition has been happening for a while. For example, every time we use autocorrect, dial a friend's number that's stored on our smartphone, or search for information on Google, a machine is being used to alter our biological capabilities. But it doesn't stop there; thousands of cyborgs walk among us every day, artificially enhanced by hearing aids, pacemakers, and prosthetic limbs. Even taking a drug to improve mental stability, get rid of allergy symptoms, or reduce the pain of a headache could be considered an upgrade. Although much of this goes unrecognized, the truth of the matter is that we've been enhancing ourselves for centuries in the pursuit of greater health, efficiency, and comfort.

However, although many of these practices are common now, they were not always accepted. At first, most new technologies are only accepted by a small group of risk takers as the majority remains hostage to old ways, an instinct learned from evolution. If everyone blindly adopted everything without question, they probably wouldn't live very long. Over time though, as new technologies become more familiar and trustworthy, the opinion of the general public tends to change. Eventually these new ways become so embedded in society that we forget what the world was like before they existed. This natural tendency to forget our old ways is the same reason future generations will look back on life today the same way we look at Neanderthals, regardless of how advanced we may consider ourselves to be.

Consequently, any fears about the speed at which modern technologies are evolving should be considered natural and reasonable. However, before overreacting, we must also keep in mind that there are a wide array of potential benefits that may come from what's happening as well. Advanced appropriately, the technologies being created can help us make an accessible, inclusive world like we've never seen before.

Voice Assistants and Brain Computing

Looking at the history of Human-Computer Interaction (HCI) models—the way we communicate with machines—it's quite impressive how much easier machines are to use today than they were in the early days of computing. The first computers required people to learn binary and communicate with the machine through punch cards. In the 1960s and 1970s, command line interfaces (CLI), which allowed people to communicate with the machine through a keyboard, made machine interactions easier, but even this was not something many people wanted to learn—most people don't want to learn a second language to communicate with humans, why would they want to learn a second language just so they could communicate with a machine?

Not long after, in the late 1970s, early 1980s, the way people interacted with machines changed once again with the rise of graphical user interfaces (GUI). Now, instead of learning a machine language, all consumers had to do was learn a system of graphical icons, which made computers much more accessible to the general public. This increased accessibility enabled computers to become mainstream consumer items for the first time in the history of computing.

More recently, natural user interfaces (NUI), which allow people to interact with machines using swipes, pinches, shakes and other gestures, have become popular as we've transitioned from desktop computers to smartphones and other forms of mobile computing. However, while this is thought by many in the tech world to be the most intuitive type of interface, many struggle to use these systems because there are no signifiers or affordances telling them what to do.

However, following improvements in the capabilities of artificial intelligence the next transition is already taking place with voice user interfaces (VUI). When perfected, voice interactions will become the most intuitive, accessible HCI model yet because consumers will no longer need to learn the language or internal system of machines in order to operate them. Instead, machines will learn our language, and our system. A good example of what this future might look like is a system developed by a team from New Zealand called Soul Machines, which uses 3D-rendered avatars capable of communicating like humans to help people navigate New Zealand's National Disability Insurance Scheme (NDIS).[1] Traditionally, navigating the system was highly difficult or impossible for many of the system's users but now, thanks to voice assistance, the interactions have become seamless.

Beyond voice computing, it is also more than likely that brain computing will become mainstream. Right now brain computing startups like Neurable are using electroencephalography (EEG) technology to enable people to operate machines with their brainwaves.[2] Often such tools are used to help people who would otherwise be unable to operate a machine, like Parkinson's disease patients, to express their full desires. As the technology evolves, it will surely be used for other purposes as well. One specific point of interest is helping people operate in virtual worlds by complementing external motion trackers, making the experience much smoother.

As we continue developing new machines to replicate different aspects of life, we will reach a point where interactions with machines will reflect all of our natural human senses. Eventually, these technologies may even allow us to experience life the way others do, which could allow us to empathize more with other people instead of remaining stuck in our personal bubbles of likes, follows, and other subscriptions. Interactions like this have the potential to bring the world together in more meaningful and harmonious ways, thanks to the help of machines.

Mechanical Enhancements

Beyond the future potential of voice assistants and brain computing, there is a world of good that is already being done with modern technologies. For example, in early 2018 a man named Johnny Matheny from Port Richey, Florida, who lost his arm to cancer in 2005, was given a robotic arm that he can operate with his mind the same way he could operate his arm before losing it.[3] Further, in warehouses and other hazardous environments, lightweight commercial exoskeletons are being used to make work safer and more accessible.[4] Moreover, in the military, similar exoskeleton technologies are enabling soldiers to catalyze their natural strength while also ensuring a level of protection that's never been realized until now.[5] These devices are enabling people to do things they were previously incapable of doing due to circumstances beyond their control or that they shouldn't have had to deal with in the first place.

The problem many companies face when trying to get the general public to adopt such technologies is that they make us "different." However, as these tools become cheaper, better designed, and less invasive, they will likely become a matter of fashion, the same way smartphones did as soon as the iPhone appeared. Ultimately, our desire to have these devices connected to us or embedded inside us will simply be a matter of how much they increase our quality of life versus how much they make us look and feel like aliens.

Universal Mobility

Combined with crowdsourced knowledge, machine intelligence also has the potential to augment and enhance mobility. The idea of getting into a self-driving car inspires fear in many people, but that's often because they lack the context as to how much more driving experience these machine have than us. Consider, for example, that the average person in the United States drives 13,476 miles a year, which equates to a little over 800,000 miles in 60 years on the road.[6] Now, compare this to Tesla's machine intelligence, which has already driven more than 7.2 billion miles, more than 9,000 times more driving experience than the average person will have in their entire lifetime, since 2003 thanks to crowdsourced driving experience.[7] According to Musk, "current fleet learning is happening at just over 3 million miles (5 million km) per day."[8] This is an incredible feat, but the system will only get better as Tesla creates cheaper cars that more people can afford, enlisting more drivers to train the system.

●   **Your lifetime of driving experience**

Contrary to what many people believe, Tesla is not the only car company working on this. Nearly every major car company has plans to have an autonomous car on the market by 2021 at the latest.[9] Despite how well these companies have disguised their intents, any car that has lane assist, self-park, back-up cameras or any other automated system has all the sensors necessary to collect training data, which means data has been collected across all terrains, weather types, population densities, and more, for years. With this data, self-driving cars will be capable of outperforming human drivers at every point of the driving experience, as demonstrated in early 2017 when a Tesla was recorded

The dots covering the last three pages represent Tesla's 15 years of driving experience, as compared to the average individual's lifetime.

saving its riders from an accident two cars ahead of it, which it predicted based on the traffic patterns surrounding it—something no human would ever be able to do.[10] Researchers believe self-driving cars will be so good at driving that they will reduce traffic fatalities by up to 90 percent, saving countries hundreds of billions of dollars over time.[11]

Consider how this could alter the future. Want to go out and party but don't want to worry about a DUI? Do it. Worried about falling asleep behind the wheel on a cross country tour? Take a nap. Don't want to park half a mile away from the venue but also don't care to spend half an hour trying to find a closer parking spot? Have your car drop you off. There are so many superficial benefits to the individual the list could go on forever. However, the real impact of self-driving cars goes beyond superficial benefits.

Self-driving cars will also enable many positive changes to the physical infrastructure of our world. For example, did you know that if you laid out all the parking spaces in the United States over one connected grid, it would amount to an area larger than Puerto Rico?[12] This is exactly what MIT Professor and author Eran Ben-Joseph, discovered in his research of urban environments. If there were no longer a need to park, this land could be used to create human-centric cities where space is used for more meaningful purposes, bringing us closer to eliminating the issues we're faced with due to urban sprawl.

What we'll see over the next 20 to 30 years is that self-driving cars will replace human-operated vehicles the same way automatic transmissions replaced the manual standard. And at a certain point, having a self-driving car will probably become a matter of public safety, causing those who choose to forgo the change to be viewed as negligent for putting the lives of others at risk. And once we reach that point, human-operated cars will be become valueless beyond nostalgic appeal, which will only reinforce the desire to upgrade.

To many people of previous generations, the idea of a world full of self-driving cars may sound like a story from *The Jetsons*, but younger generations are already prepared for it. Thanks to ride-sharing apps, kids today are waiting much longer to get their license, some are even forgoing the process completely.[13] As ride-sharing services become fully automated, they will only get cheaper and more efficient, which will make the decision to forgo a driver's license even easier—something that will benefit many, beyond the youngest generation. For example, at 87 my grandpa was not allowed to drive because his poor eyesight made him a danger to others on the road. To my grandpa (and many older people in his situation) losing his keys was more than the loss of a car, it was a loss of freedom; a sentiment shared by many, of all ages and ability levels. With access to self-driving cars, things will be different. Eventually, self-driving cars will flow through city streets like water through a river, enabling a form of mobility unlike anything we have ever experienced before. As we progress, more of our natural biological abilities will be replicated into machines as well, enabling us to create a digitally enhanced biosphere that amplifies the best of our natural world, allowing it to communicate, respond, and cooperate like us.

The Kids Don't Care to Drive

Percentage of 12th graders who drove at all in the last year and who have a driver's license.[14]

- ■ Have driver's license
- Drove at all in last year

Percentage of Respondents

Years, 1976–2015

CHAPTER 13

Ambient Machines

Once our natural abilities are replicated and turned into machines, we will be capable of extending them to every piece of life. This will allow us to create a world in which our environments reflect the nature of our humanity—a world in which cars, buildings, and entire ecosystems are capable of co-existing with us, just like the natural world.

As the world around us becomes seamlessly connected to the internet, we have the potential for an even higher level of experience, enabled by an intelligent network of sensors. Imagine a not-so-distant future when buildings are capable of monitoring the weather, adjusting electrical outputs and water flows to optimize efficiency, giving homeowners everything they want while decreasing consumption to zero. With systems like this we wouldn't have to work so hard to be sustainable, the system would do it for us. The idea of such systems being real reflects Arthur C. Clarke's statement that, "Any sufficiently advanced technology is indistinguishable from magic."[1] Yet, these developments are already becoming a reality for many people thanks to smart home devices like the Nest Thermostat, Phillips Hue lightbulbs, and other intelligent home products.

Such tools are an incredible start, but imagine when companies start embedding them throughout office spaces, like the Bullitt Center in Seattle. Considered one of the greenest buildings in the world, the building operates at a net-zero energy level by generating all of its energy from solar panels and capturing water from rain, which it then cleans so that it is consumable by occupants. Light and temperature sensors monitor daylight and heat, and communicate with other connected devices in the space to adjust window shades and light intensity based on room occupancy, external weather, solar radiation, and more. These interactions regulate internal temperatures and electrical usage in a way that enables the building to use an astonishingly low amount of energy.[2] Connected buildings and ecosystems that understand their surroundings and internal usage are not just fancy upgrades, but creative responses to the needs of the world.

Beyond buildings, imagine a world in which entire cities come to life through smart streetlights capable of monitoring traffic, enabling better traffic flow; audio sensors that monitor noise levels and maintain low levels of noise pollution; and Wi-Fi that flows like the air we breathe, enabling everyone to have access to internet as if it were a basic utility. Again, although this may sound like science fiction, it's already happening in many cities around the world, including Los Angeles, which has invested $57 million to convert 215,000 streetlights to smart lampposts with LEDs. This will not only give the city better control over the lights but also permit them to use lights that last longer, which is expected to save LA more than $53 million a year, meaning the city will recoup its costs in just 13 months.

Beyond the cost savings created long term, this system will also offer incredible benefits by turning the city into a connected platform that monitors sound, motion, weather, smoke and fire, open parking spaces, and just about whatever else citizens may need in order to optimize their experience. Similar projects are underway in other cities around the world, including San Jose, San Diego, Chattanooga, Newark, Detroit, Buenos Aires, Madrid, and Copenhagen.[3]

Another critically important part of our cities that is in need of an upgrade is waste management. Current systems lag well behind what is possible today, but imagine a world in which smart trash cans are capable of monitoring themselves and helping public service

teams make their job more efficient. Thanks to a company called Big-Belly Solar, which has created smart trash cans that are now installed across several US cities like Boston, New York, and Philadelphia, this too is a reality. These bins have been designed not only to monitor the levels of trash within themselves, but also act as trash compactors, which means they don't have to be emptied as often—all of which is powered by solar energy. And, because they're connected to the Internet, these bins are capable of signaling to waste management teams, which simplifies logistics and lowers CO_2 emissions by only routing trucks to full trash cans. This means cities can save the environment while saving time and money.[4]

The hurdle to creating city-wide change, or greater, is often the price tag that comes with such upgrades. Convincing people that spending millions of dollars on "smart" lightbulbs or other connected devices that operate invisibly, as ambient technologies, is often hard because most of the tools remain unfamiliar and untested. But even if we think on a smaller scale, reducing the risk by and cost of investment by using things that already exist and are connected to the internet, there are plenty of benefits to be discovered through simple, little connections without city-, state-, or nation-wide infrastructure needing to be installed. Consider, for example, that at this point in history almost every car on the road is connected to the internet. Knowing this, we could create a system that monitors changes to each car's tire pressure and suspension to recognize when drivers hit a pothole. This system could then send the GPS location of each pothole to a database that stores where they're located and how often they're hit, which would enable the city to both locate potholes and prioritize which get filled first. This is a small idea that wouldn't cost much to get started and could very easily grow into something much larger, over time.

Imagine longer term, once we have robots capable of filling potholes, a world in which we could use such a database to send robots out to automatically fill holes instead of having to send out a team of humans who have to risk their lives working on busy streets and highways with nothing but cones protecting them from speeding cars. And because robots don't need sleep, this work could be done while the average citizen is sleeping, enabling operations to be performed with greater speed and reducing the amount of construction people have to suffer through on their daily commute.

What the Future Could Be

What we've just covered are only a few examples of what is possible right now. As this revolution continues only time will tell exactly how things will manifest. However, consider that we already have autonomous ships skimming the ocean to clean up oil spills; zero-gravity 3D printers capable of operating in space to send tools and parts up to astronauts who no longer have to wait for supplies to be shipped from Earth; and trees being turned into antennas capable of communicating over long distances.[5][6][7] Ideas like these are just the beginning though.

Sensors embedded in everything we own are just waiting to be used to their full potential.

As we move forward and more items are connected to the internet, these sensors will become what amounts to a central nervous system for our inorganic world. This will enable our world—once only static, like a rock—to observe, engage, and heal just like every other natural organism we coexist with. Used ethically, this will allow us to maintain our world in much more effective, sustainable ways while enabling people to focus on the more meaningful aspects of life. The benefits these technologies are capable of bringing to our world are why I, as well as many other creative technologists, got into the field. However, while these technologies have incredible potential to improve our world, there's definitely still plenty to be concerned about as well. And until we fix the current problems underlying the internet, making meaningful change will be difficult.

PART 4

How Do We Get There?

Despite the threats created by modern technologies, plenty of good has been done as well, which should not be understated or down-played. However, if we want to make the future enjoyable, we must discover ways to amplify the good parts and curb the bad.

CHAPTER 14

Recognize Data Monopolies

Antitrust laws tend to be utilized as soon as they become a matter of public safety. In an analysis of the current structure of the market, it is clear to see that a handful of companies are now controlling the market on a global scale and this is beginning to impact the public's safety. Making changes to this situation is something many people have begun to demand, but where do we start?

From the time of Northern Securities in 1904 to Standard Oil in 1911, all the way to American Telephone and Telegraph (AT&T) in 1982, antitrust laws have been implemented to protect consumers from the manipulation of marketplace owners for more than 100 years. More recently, there has been concern that Big Tech is reaching the point of monopoly control, leading many people to call for antitrust protection. But what exactly is antitrust law? And why is it implemented? Well, according to the FTC, antitrust laws, "protect the process of competition for the benefit of consumers, making sure there are strong incentives for businesses to operate efficiently, keep prices down, and keep quality high."[1] However, Tim Wu, in his book *The Master Switch: The Rise and Fall of Information Empires*, points out some problems in calculating the efficiency of such practices:

> *As I have argued repeatedly, there are issues in the information industries that render the traditional efficiency calculations informing most regulatory policy inadequate. The most glaring deficiency is the habit among antitrust practitioners to insist on identifying failures of competition entirely by looking for an effect on prices (the problem of "price fixation"). In practice, however, not all pernicious combinations inflate costs to the consumer.*[2]

Despite the fact that there are three main pillars to antitrust law: price, fair trade, and fair competition, there is one (price) that tends to outweigh them all. If regulators discover that a company is manipulating the process of free trade or completely destroying competition but is actively lowering prices in the market, they will generally let the monopoly continue. This makes implementing antitrust law exceedingly difficult in the age of the Internet, when most products we use are free in the traditional monetary sense. That being said, the new marketplace we exist in, where we pay with our attention, has thrown regulators for a loop, and for that reason, we'll start there.

Price Gouging

In order to identify price gouging in this new marketplace, we need to better understand attention, the psychological currency mining the life out of consumers. As a metric, attention is typically measured by analyzing two major factors: (1) number of users and (2) amount of time spent using the product—or "engagement." The goal of companies in this new attention economy is to acquire as many users as possible and to keep them engaged as long as possible by making what the industry calls a "sticky" experience. Success is then measured as monthly active users (MAUs) and Time on Site (TOS), which are considered numbers that companies can show investors and shareholders to indicate how well the platform is performing. Why? Because when consumers pay more attention any given application, their interactions turn into data, which then gets turned into money.

The problem with the desire to acquire more data is that products are now being intentionally created to release chemicals in our bodies the same way drugs and other addictive experiences do, to drive

engagement as high as possible. Thus, traditional fiscal costs may be at or nearing zero, but the cost to the world's mental and physical well-ness is skyrocketing at an alarming rate. With this understanding we can reframe the conversation, and better recognize price gouging in the attention economy.

To start this conversation we first have to recognize one major flaw in the system, which is that currently, attention is an illegitimate and corrupted currency. While attention is bought and sold in digital marketplaces in the form of data, there's no denying that it lacks a tangible value—it's not like gasoline that costs $2.75 a gallon, or a dollar bill, which can be passed around without losing its value. On top of this, attention is also corrupted in the sense that the practice of capturing it is damaging communities around the world both mentally and physically. Although I, as well as many other experts, believe there is inherent value in attention and the data created by it, lacking a publicly traded value allows it to be easily manipulated with relatively little risk of getting caught. With this understanding, we can see that what's really happening behind the scenes is not so different from money laundering. The way the system currently works, companies can take an illegitimate (or "dirty") currency in attention that most people are incapable of recognizing, wash it through a legitimate, public-facing business model of data brokering and product creation, and turn it into a legitimate currency that can be used in other, legitimate markets. This isn't to say we should investigate these companies for money laundering, rather it is to critically assess the industry and shine a light on the fact that these companies have created a black market industry worth trillions of dollars with almost zero tangible production value.

What's happening today is similar to what we saw in the financial industry in the early 2000s when companies complexified the industry to the point outsiders were incapable of understanding it, then leveraged their enhanced literacy and insider information to manipulate the market to their advantage. This allowed the financial market to skyrocket for a brief period before crashing down thanks to the compounding effects of corruption, leading to the largest financial crisis the world has seen since the Great Depression. This also explains why Big Tech companies have recently been seeing 40 to 60 percent returns each year on the stock market. A lot of these issues could perhaps be resolved if there were a consensus on the value of data and attention, but this currently is not the case.

Finally, even if attention were a legitimate currency there's one last problem that often goes unnoticed: by acquiring as many users as possible and driving engagement rates through the roof, all companies involved are actively and intentionally price gouging. Although it is not easy to understand because of how abstract this marketplace is, this needs to change. Like traditional monopolies, which were allowed to continue operations if they were lowering fiscal costs, modern data monopolies should only be allowed to proceed if they are working to lower attention costs. Finding a price point that enables businesses to thrive without sucking the life out of consumers will require considerable work, but is necessary if we hope to resolve these issues.

Elimination of Free Trade

Antitrust law is also dependent on compelling evidence of unreasonable restraint of free trade—whether through acquisition, merger, or sheer dominance—which traditionally meant, "arrangements among competing individuals or businesses to fix prices, divide markets, or rig bids," leading to "higher prices, fewer or lower-quality goods or services, or less innovation."[34] The question that remains is: *With data replacing physical goods as the new commodity being traded, how should regulators prove this*? The answer is that modern interpretations should focus less on the free trade of goods and more on free trade of data. As the European Parliament has written in Europe's General Data Protection Regulation (GDPR), "Flows of personal data to and from countries outside the Union and international organizations are necessary for the expansion of international trade and international cooperation."[5]

At the moment, it is common practice for platforms to make what the industry calls "seamlessly digital ecosystems" that work so well together that strategic value is added to the product ecosystem—buy one Apple product, you buy them all because of how well they work together. Building hardware platforms that work together seamlessly is brilliant business strategy; data can easily move between internet connected devices, so it doesn't matter.

However, when software platforms do this, companies are essentially creating what amounts to a data corral. For example, Facebook (as well as many other platforms) makes sharing between platforms difficult in order to keep data within its ecosystem; in 2009, Apple strategically decided to keep Google Voice out of its App Store, forcing consumers to use Apple services; and for the past decade, Google has been deranking companies like Foundem, a search competitor, because losing consumers to a different, potentially better search engine, would be a liability to their data collection practices.[67] Such behaviors may seem petty to consumers who don't understand the value of data exchanges, but to the businesses involved they represent a strategy to contain users and aggregate as much data as possible.

In defense of this behavior, the American Bar Association (ABA) made a statement in early 2018 arguing that the value of proprietary datasets is not what is prohibiting businesses from competing: "While customer data is an important input for businesses, it is often replicable or substitutable, because one firm's collection of data on customers does not preclude another firm from collecting the same or similar data on customers from other sources...the existence of large proprietary databases is not necessarily indicative of dominance."[8] While the Association is correct that the existence of large datasets is not necessarily indicative of dominance, it would be irresponsible to believe that when leveraged appropriately these assets do not create a moat that enables platform owners to stifle competition. The ABA clearly demonstrates its lack of coherent understanding by continuing on to state: "The Sections respectfully submit that the possession of data advantages could be a potential source of market power only if, at the very least, the data were necessary to compete, served as an effective barrier to entry, and comparable data were not replicable or accessible by other companies in a timely manner,"[10] which is the real issue.

The reason these companies create seamless digital ecosystems is partly because it makes for a better customer experience, but primarily because it contains data within their platform, giving them a decisive strategic advantage over the competition. Sure, many companies have free APIs, but what often gets lost in translation is that they also have complete control of these APIs, and once a company decides maintaining an API is no longer what's best for business the service often becomes unmaintained and unusable—or worse, deprecated completely without any meaningful alternative. This leverage enables companies to destroy entire marketplaces with the flip of a switch.

Destruction of Competition

On top of being able to control the flow of data, a company that owns a data monopoly on any single vertical or combination of verticals can then use that data to refine its products in a way no competitor would ever be able to. It can then leverage this performance advantage to persuade more consumers to use its product. The exponential growth of the platform that results from this then creates an impenetrable network effect through Metcalfe's Law, which then combines with an exponential increase in data to synergistically dominate and suffocate all competition.

This brings us to the last major pillar of antitrust law — destruction of competition. In the digital world, destruction of competition should be less about whether competition can literally be created and more about whether the company has gained an unreasonable strategic advantage through its data assets. Today competition is restricted less by the ability to create the platform and more by the platform owner's data assets and network of users, who then feed the company data. In the tech world anything can be created. Creating a business is cheaper, faster, and easier to iterate on than ever before. But just because it can be made doesn't mean it can exist on its own.

For example, Facebook owns four of the top seven social media platforms in the world.[11] If you remove WeChat and QQ, which are made to keep Facebook out of China and generally not used by the rest of the world, Facebook then owns four of the top five social media platforms in the world. Because of this, Facebook has exclusive access to more data and users (see, attention) than any other social media company in the world. They can then use their assets to create a product no competitor would have the insights to understand how to build. And by leveraging the power of Metcalfe's Law, they can then force competitors into a position where they're forced to decide between two options: (1) agree to be acquired or (2) take on Facebook head-on.

We can also see this when we look at the performance advantage Google owns over Amazon in voice technologies. In a study by 360i, it was determined after asking each device some 3,000 questions that Google Home was six times more likely to answer your question.[12] Amazon performed relatively well when it came to finding products, but in the overall performance of search and user queries, Google dominated. This happened despite the fact that at the time the test was run Amazon owned more than 70 percent of the connected speaker market and had released Alexa to the public more than two years before Google Home.

This, in large part, has to do with the monopoly Google owns on search data. According to reports, Google owns over 73 percent of internet search data worldwide.[13] This is an eye-opening statistic considering Google services are completely blocked in China, which represents more than 18 percent of the global population. If we considered how much Google owns without China, it could be argued that Google actually owns up to 90 percent of search traffic worldwide. And in fact, this is exactly the percentage of the European search market Google owns, according to the EU's antitrust suit against the company.[14]

Surely part of each company's dominance has to do with the way their system is designed and engineered, but a large part of it comes down to data. Owning a data monopoly is why Google Home is superior in knowledge queries when compared to any other speaker, why Amazon is better at delivering retail search results than any other retailer, and why Facebook is capable of understanding your social interactions better than any other platform. And this is why many other businesses are in pursuit of monopolizing their vertical as well. This level of control gives the platform owner a strategic performance advantage within its products, which then gets delivered to the end consumer, which then drives higher adoption and engagement rates, which then allows the company to collect more data and synergistically catalyze their monopolistic pursuit. It's a vicious cycle that underlies the cancerous growth of the internet and has left consumers with no alternative to the platforms that currently exist.

Allowing these corporations to dominate the world without restraint has led to a state of the market where 53 percent of startups in Silicon Valley believe the long-term goal of their company is to get acquired by one of the giants. Only 18 percent believe they have a chance at remaining private, and only 16 percent believe they stand a chance at going public independently. This is because these entrepreneurs recognize that they will never have the money or the power to put them in a position to challenge these giants alone. What's worse? More than 50 percent surveyed believe this will only get worse as time goes on.[15]

At this point in history these organizations are no longer innovating internally. Instead, they are innovating through acquisition, leveraging their monopoly power to force competition to give in or face complete destruction. And because of this, we're seeing less companies formed and publicly offered than we've seen in the past, a fact that is clearly demonstrated in a study by the National Bureau of Economic Research, which found that the total number of public companies in the United States has fallen by almost half since 1996—which is exactly when commercial Internet companies really started to take off.[16] This is a trend that is bound to continue if we don't do something.

Time for Change

The purpose of antitrust law, as stated by the United States Supreme Court, is, "Not to protect businesses from the working of the market; it is to protect the public from the failure of the market. The law directs itself not against conduct which is competitive, even severely so, but against conduct which unfairly tends to destroy competition itself."[17] In recognizing the intangible differences between today's monopolies and the monopolies of yesteryear, we must also recognize that the way antitrust law is implemented will surely be different.

Reorganizing modern tech companies will have less to do with their product ecosystem and more to do with their data collection strategies. For example, Amazon works hard to strategically monopolize retail and logistics data; Facebook, social and emotional data; and Google, knowledge, in general. These companies do not represent a conclusive list—many others have their own—but these companies do make it easier to understand the issues because of their familiarity. Only by analyzing a company's data acquisition strategy can we begin to understand what the organization is attempting to monopolize and discover what needs to be done to ease the company's grip. After proper analysis and consideration, it is imperative that regulators implement the appropriate measures to turn things around, not only to assure freedom of competition, but also to ensure a reasonable level of public safety. When consumers have no viable alternative, it does not mean they consent to usage, but rather that they have no other choice: the definition of monopoly.

CHAPTER 15

Make News Trustworthy

We've reached a point in history where citizens are incapable of trusting their own two eyes. It has put the population into a position where they are forced to debate the meaning of truth. Having conversations about how we fix this situation will be incredibly intimate and uncomfortable, but they are discussions we need to have.

15

In the United States alone more than two-thirds of the population get their news from social media, as do a similar amount of Europeans, according to doteveryone, a digital think tank from the United Kingdom.[1][2] Even those who don't are influenced by this phenomena, as social media posts infest their news broadcasts, print publications, and elsewhere. This is a large percentage of people getting their news from what are currently recognized as entertainment platforms—which has made judging the accuracy of news increasingly difficult and led to troubling repercussions. As Chamath Palihapitiya, former vice president of User Growth at Facebook has said:

> We have created tools that are ripping apart the social fabric of how society works... The short-term, dopamine-driven feedback loops we've created [Likes, Hearts, Thumbs up, etc.] are destroying how society works. No civil discourse, no cooperation; misinformation, mistruth. And it's not an American problem—this is not about Russians ads. This is a global problem.[3]

Furthermore, it has pushed social media companies out of the realm of fun, social entertainment platforms, into the world of news and information distribution, putting them in a powerful position to control the world's perception of reality. These companies have grown so big and influential that they could be considered the intellectual infrastructure for billions of people across the globe, as Mark Zuckerberg has said about Facebook:

> I think there's confusion around what the point of social networks is. A lot of different companies characterized as social networks have different goals—some serve the function of business networking, some are media portals. What we're trying to do is just make it really efficient for people to communicate, get information and share information. We always try to emphasize the utility component.[4]

In pursuing this goal, these organizations have silently become the largest news distribution organizations in history, which makes people feel like everything they see is their choice, and is much more difficult than identify than centralized content sources where people can easily recognize who is in charge. Yet when it comes time to discuss their role in modern news curation, platform owners continue to downplay their role because embracing it would require them to figure out ways to ethically curate news, establish editorial objectivity, and reliably fact-check at the scale of the internet—an incredibly difficult task that would surely negatively impact their bottom line. Eventually though, changes must be made. What we need now is a proactive, adaptable system that allows citizens to trust their news again, of which, the fundamental principles must include reducing dangerous bot populations, defining what it means to be an "expert," and adding more humans to the system to evaluate content.

Reduce Dangerous Bot Populations

As technologists it's generally understood that bots can be used for good or bad. However, just like bad actors and criminals in real life, it's the bad bots that stain the reputation for the general public. From manipulating stock markets to persuading individuals into poor decisions, to manipulating elections and more, the public has come to recognize bots as one of the most dangerous parts of the internet—and rightfully so considering bots made up more than 51 percent of internet traffic worldwide in 2016, more than 56 percent of which were used for malicious purposes.[5][6][7][8][9][10] Despite this, we also recognize that the internet would not be manageable without them.

From Google Search, which helps us find information, to customer service agents that expedite service, to fraud protection providers that protect our most valuable assets, there is an incredible amount of good that comes from bots. Unfortunately, with the way the internet works at this point in history there are loads of financial incentives for, and very few reasons to stop, making bad ones. As Dr. Angelos Keromytis of the United States Defense Advanced Research Projects Agency (DARPA) explains: "Malicious actors in cyberspace currently operate with little fear of being caught due to the fact that it is extremely difficult, in some cases perhaps even impossible, to reliably and confidently attribute actions in cyberspace to individuals."[11]

To increase the friction and reduce the number of bad bots, many have begun to call for greater verification—through Social Security numbers, biometrics, and a broad combination of various other sensitive details. While these sound rational, we need to be careful what we wish for. For one, the more sensitive the information required to verify, the more companies must to be trusted to safely process and secure our data. This leads to greater security threats on both sides—people have to trust the company won't be hacked and companies have to be more concerned about hackers with greater incentive. Beyond security, collecting more data also means more storage space, which means greater cost. Lastly, legally acquiring this data means navigating the nuances of governments around the world. For these reasons, asking for more sensitive data is not the smartest strategy for consumers or businesses, especially at a globe scale.

Second, despite how safe and efficient they may appear, biometric verification through fingerprint scanners, Face ID, VoicePrint, and other technologies can all be hacked just like any other password. A hacker group called the Chaos Computer Club hacked Apple's fingerprint scanner with simple household tools.[12] World famous hacker Jan Krissler used high resolution photos to recreate the fingerprint of Germany's Prime Minister of Defense, Ursula von der Leyen.[13] And still others have broken into the iris recognition system on Samsung's S8.[14] What these cases prove is that while biometric identification may currently be more effective than a password, they're not impenetrable. The difference with biometric identifiers is that they can't be reset if they get stolen, unlike a password.

So how do we make sure the people behind the screen are real humans? Is it even possible on a global scale? A good example of how this might happen is what Google does to verify business owners—send a verification code in the mail. This is a tactic that operates around the globe, doesn't require highly sensitive information, and can be done by almost anyone. While some may argue that not everyone has an address, that's quickly changing with systems like What3Words, which turns the entire world into a digital grid.[15] If nothing else works, people can also phone in, which more than two-thirds of the world is capable of doing.[16] By rooting the authentication in the real world we bring a traditional process into the digital world and create verification that surpasses digital security.

Define What Being an Expert Means

Once we can verify the humans behind the bots we can then begin to establish expertise, which, unfortunately, tends to have a lot to do with the size of someone's follower count, regardless of whether it's real or not. This is a flaw of the herd mentality we possess as human beings. However, the danger in this is that there are many simple strategies for faking it, and because of this, we now have more "experts" than ever before—many of whom have no expertise at all. A major flaw in this system is that there is a lost of economic incentive to manipulate and inflate these numbers because they are considered key indicators of success, despite being relatively valueless beyond face value. This could be changed by creating a system that identifies various levels of verified experts—perhaps local, national, and world, for example; requires experts to prove their expertise in standardized ways; and requires a fee. Carefully considered, a solution such as this would enable platforms to make change without re-architecting the entire platform, meaning changes could be implemented quickly.

To validate higher levels of expertise, platform owners could require training courses that ensure a baseline understanding of the ethical responsibilities that come with influencing the world, something traditional mass media experts had to do in college or otherwise learn on the job, from those who had been trained. This would also offer a second layer of verification to ensure accounts are not bots, which would enable us to separate real, human experts, from malicious actors looking for the quickest way to increase their influence with the least amount of effort.

To maintain fairness and objectivity, requirements should be mutually agreed upon by a regulatory body with public ties—whether that be a traditional government agency, independent civic organization, or both—not privately held corporations. These companies should, however, be involved because their years' of experience with these issues will add value to the discussion. Once defined, these rules should then be displayed transparently so anyone can understand what being an "expert" means and scrutinize the process. Some dangers that may result from such a system include eliminating opportunity for (1) those who cannot afford to pay and (2) businesses and industries that are

1 The way verification works on most platforms today, is that anyone can email the platform asking to be verified. The decision of who gets verified and who doesn't is then determined on a subjective basis by those in charge of handing out verified badges.

This means we have to trust those in charge with blind faith, because there is currently no way to scrutinize their decision.

non-standardizable, such as music and art—which we'll need to avoid.

The fees from this process—which would force those gaining the most value from the platform to fund it—would also make it so companies wouldn't have to rely on advertising as their only revenue stream. As a reward, the people who pass this test could perhaps get their posts weighted higher in the algorithm or receive free or reduced-cost ad space to promote their work—either of which would allow greater influence by those who have proven their trustworthiness, without censoring or reducing the voices of others. There's an obvious counter-argument to this, which is that this would create a position of privilege on the internet, which is not entirely wrong, but the same argument could be made for doctors, lawyers, personal trainers, hairstylists, taxi drivers, and many other professionals who have to obtain a license to operate—even dog walkers are required to certify themselves in certain states. Such licenses are not created to put anyone in a privileged position but to ensure a level of public safety, something we desperately need on the internet. Obviously a certain demographic of bad actors would still be able and willing to operate within this system, but as with any law it is not about completely eliminating bad actors, it's about reducing the ratio.

2 Getting away from an advertising-only revenue model would allow companies to focus on creating products that serve consumers' needs instead of serving them ads.

Require More Information Respondents

This brings us to our last, most sensitive topic, which is how to ensure information is valid without censoring anyone or suppressing freedom of speech—a debate that will likely be fought for the rest of our lives. Currently, most companies are working on algorithms to audit the content on their platform. From the outside, this appears to be a suitable response that removes subjectivity and bias but in reality this is just reinforcing Big Tech's control of the internet. However, this push for algorithmic supervision instead of humans will continue for a couple of reasons. First, it is true that we need automated systems to help us discover what needs to be looked at, which makes it hard for those who don't understand the issue to debate it. However, while machines are good at large-scale aggregation, they still need humans to assist them in analyzing the nuances. As Dr. Jennifer Roberts of DARPA has stated:

> *Enterprise-sized networks present challenges in terms of both their size and distributed structure. Today's state-of-the-art commercial tools do not directly address the scale and speed needed to provide the best defense for multiple networks... Current commercially available tools may output thousands of alerts and false positives per day that often cannot be verified due to a lack of processing capacity... Real-time detection of threats within or across very large enterprise networks is not simply an issue of scale, but also a challenge due to the variable nature of malicious activities and their presentations.*[17]

Despite the fact that several companies have committed to hiring anywhere from 10,000 to 20,000 additional workers, this is nothing compared to what's needed.[18] Consider, for example, that in New York

City alone there are more than 15,000 emergency responders assisting nine million people, many of whom they struggle to assist meaningfully because there are limitations to time and space. Now, compare those numbers to Facebook, which has offered to hire 20,000 additional respondents to take care of more than 2.1 billion people.[19] That means Facebook has offered to hire 1.3 times the amount of workers to assist more than 230 times the amount of people. What should be clear, given context, is that we need more humans in the system—a strategy that, as of April 3rd, 2018 has been adopted by DARPA.

In order to enhance cybersecurity, DARPA, a research arm of the United States military created to spur technological innovation, has created a program called Computers and Humans Exploring Software Security (CHESS) to make their systems more resilient against cyber attacks. According to an article from Defense One, DARPA believes that "while computers outshine humans at spotting vulnerabilities and repelling attacks that mirror basic logic and math problems, humans remain better at problems that follow a more complex set of rules."[20] Moreover, a press release from DARPA mentions that, "moving from a manual, human-driven process to one that is based on advanced computer-human collaboration creates opportunities for a broader range of technical—or potentially non-technical—experts to assist in the detection and remediation of known and emerging threats... By allowing more individuals to contribute to the process, we're creating a way to scale vulnerability detection well beyond its current limits."[21]

Despite the clear need for more humans, Big Tech companies avoid hiring humans because it's cheaper to use automated systems than it is to hire the hundreds of thousands (or millions) of people necessary to address the current crisis properly. But cost is not a valid argument. These companies have more money than many nations combined, and this should be used to address these issues accordingly. Future legislation should consider measures that require every content distribution platform on the internet—current and future—to have a certain amount of information respondents on staff, relative to the number of flagged posts received each day. This would be good both for public safety and job creation.

For example, older, experienced doctors who are ready to retire but don't want to completely disengage with the medical field could contract part-time to help correct or verify medical information. This would give them the flexibility to work wherever and whenever they want, while remaining engaged with the field by leveraging their experience to make medical information more trustworthy. The crisis experience police officers and other emergency respondents possess could be leveraged to begin creating a 911 system for the web, which is something we currently lack. Journalists could help fact check and edit, acting as editorial gatekeepers for the internet—a concept that could even become a revenue stream for platforms. These are only a few examples, but this same thinking could be extended to every individual on the planet, which would create a renewed value for millions of people in the age of automation while simultaneously making the internet trustworthy.

News We Can Trust

After reading this chapter you're probably filled with a lot of sensitive thoughts and feelings, but that's exactly the point. Life's not all rainbows and butterflies, despite what our filter bubbles have convinced us of in recent years. Anything that creates this much animosity just thinking about is something worth talking about as well. Sitting on our hands hoping it plays out safely because we're too afraid to approach these topics head on is just as, if not more, irresponsible than doing nothing at all.

What we need to focus on moving forward is creating proactive systems that add friction to the process and make it harder for bad information to get online in the first place. This isn't about eliminating the freedom of speech or the freedom to share information, it's about reducing the incentives and opportunities available to malicious actors interested in manipulating the system. To make this happen these systems need to have roots in our physical reality and should be complemented by humans. Solving these issues will be no small task, but done well, we have the opportunity make the internet safer and more democratic than ever before.

CHAPTER 16

Demand Transparency

Algorithms are now shaping the world around us. They help us discover items we might like to buy, shows we might like to watch, and restaurants we might like to eat at, among other things. Without algorithms, the internet would be unmanageable, but without a system of checks and balances, they can also become very dangerous.

In summer 2014, the State Council of China released a document detailing a radical new idea officials had been working on called a "social credit system," that would change life for the citizens of China—a plan that has since come to life and is currently operating. Today, Chinese citizens are monitored to assess their value to the state based on everything from what they buy, both online and not; where they live and travel; what they say online and how it's said; who their friends are and what they do together; how much time they spend surfing the internet and what they click on; what bills they do or do not pay and whether their payments are on time; and more.[1] Although participation is optional today, it will become mandatory in 2020.

This score will impact everything from the citizen's eligibility for jobs and financial support to their ability to get a date, enter a restaurant, and use public transportation. It will even determine what school their children are allowed to attend. Worse yet, everyone's score will be publicly available, creating a nation-wide social ranking—a modern day scarlet letter, if you will. As ominous as this sounds, the Chinese government believes such a system will enhance trust nationwide and will promote a society of sincerity.[2] This might be true if life were a perfect utopia where mistakes never happened but, as Rachel Botsman, mentions in her book *Who Can You Trust: How Technology Brought Us Together—and Why It Could Drive Us Apart*:

> *Where these systems really descend into nightmarish territory is that the trust algorithms used are unfairly reductive. They don't take into account context. For instance, one person might miss paying a bill or a fine because they were in hospital; another may simply be a freeloader. And therein lies the challenge facing all of us in the digital world, and not just the Chinese. If life-determining algorithms are here to stay, we need to figure out how they can embrace the nuances, inconsistencies and contradictions inherent in human beings and how they can reflect real life.*[3]

The idea of a social credit score probably makes your skin itch, but realistically many developed nations are not far off. Today, many of us already rate each other on Uber, Airbnb, Yelp, and other applications, creating a public facing social ranking; Fitbit, MyFitnessPal, and other tools already push us towards our "best" selves and are slowly beginning to be tied to our healthcare benefits; we can already block individuals on our social feeds if we don't want them to exist anymore; and, more generally, everything that gets surfaced to us through algorithms determine our perception of the world. Without algorithms the web would be unmanageable and therefore unusable. However, it's time we begin to recognize the power within these systems and treat the logical rules that define them as policy in code to ensure a level of safety.

Understand the Limitations of Data

Although they now govern the modern world, algorithms are only as unbiased and rational as the humans that create them and the data that flows through them, which has led to a fundamental flaw in the systems underlying the internet—the immense amount of data these systems are lacking. Although billions of people are connected to the Internet, many do not actively participate. Many of the most influential people in the world are too busy for the internet; many of the people creating these tools avoid them because they know how easy it is to get lost, billions of others just don't care to participate.

Nobody should be required to participate on the internet, but because of this lack of participation, the internet is now skewed towards the people with the most freedom and desire to participate, who often turn out to be the loudest assholes in the room—a point that was proven by the evolution of Microsoft Tay, a chatbot that turned into a belligerent asshole within 24 hours of being trained by people on Twitter.[4] Poor statistical representation has already been clearly demonstrated in algorithms that mistake African Americans for gorillas, that recognize men's voices better than women's, and that simply fail to recognize the difference between a chihuahua and a blueberry muffin.[5][6][7] The point here, as Friedrich Hayek points out in *Individualism and Economic Order*, is that data sets have limitations:

> **"** *If we possess all the relevant information, if we can start out from a given system of preferences and if we command complete knowledge of available means, the problem which remains is purely one of logic... This, however, is emphatically not the economic problem which society faces... The data from which the economic calculus starts are never for the whole society.*[8]

This is why it's important that we not only make it easy to contribute data and give feedback, but also reward people for meaningful contributions. Google Maps does this by giving people points, badges, and access to beta releases; Facebook does similar things; others are even attempting to allow people to monetize their data for charity.[9] These are good attempts, but these systems clearly benefit the recipient much more than the contributor. Eventually, rewards will have to be greater, especially as jobs are removed by the systems these contributors are training. Whether that reward is monetary in the traditional sense or something different has yet to be determined, but is something scientists are working on.[10] Once a worthy solution is discovered, the real internet revolution can begin.

Negate Proxy Damage

Good sources of data and statistical rigor are the foundation of healthy automation, which means after we figure out a way to improve data collection, we'll also need to be careful about what gets put into code. For example, algorithms work well with quantified parts of life—things that can be represented by numbers—like sports. Sure, analyses may differ from team to team, which may result in different outcomes, but, in general, anyone can get access to the data, which makes it scrutinizable and ensures nobody can manipulate the data to their advantage since stats are final at the end of the game. Thus sports analysis represents a high level of statistical rigor and algorithmic transparency. The same cannot, however, be said for every part of life, which leads to "proxies," or approximations, that significantly decrease the statistical rigor of any scientific analysis.

Undeterred by this fact, Big Tech continues to press forward in its attempt to discover the best way to quantitatively interpret qualitative aspects of life. This has led to a world where we're removing the human from the equation, as Yuval Noah Harari mentions in his book, *Sapiens: A Brief History of Humankind*: "Our computers have trouble understanding how Homo Sapiens talk, feel and dream. So we are teaching Homo Sapiens to talk, feel and dream in the language of numbers, which can be understood by computers."[11]

The thing is, if there were reliable performance metrics for love, happiness, satisfaction, and excitement, anyone should be able to perpetually satisfy these feelings as long as they were willing to put in the work. Although Big Tech would like to think otherwise, there is no code that can accurately define these experiences in a binary, or even multivariate, outcome. What's fun to us may be boring to others despite similar interests. What was heavy to us might be light to someone with a similar body type and fitness level. One Italian might like the taste of pasta, while the other can't stand it. Life exists on a continuum and these issues cannot be resolved with equations.

Create Explainable Artificial Intelligence (XAI)

In addition to recognizing the limitations of data and proxy values, much work will also be needed to make algorithms more transparent to the public, especially if these systems are to be trusted. When machines are positioned as experts and perform at a level that is intelligent enough to pass, the public will generally trust them until they fail. However, if machines fail in ways the public is incapable of understanding, they will remain expert despite their failure, which is one of the greatest threats to humanity. Consider, for example, if someone were to use a visual search tool to identify the difference between an edible mushroom and a poisonous mushroom and the machine told them a poisonous mushroom was safe. That person could die. Or what about when a machine is given the power to determine the outcome of a court case but doesn't provide an explanation for its decision? This already happens.[12] Worse yet, how about when these tools are given the right to use lethal force for military purposes? The United Nations is actively debating this issue at the time of this writing.[12][13]

To ensure the public is capable of understanding what's happening behind the scenes we need to create explainable artificial intelligence (XAI)—tools that explain how machines make their decisions and the accuracy with which these tasks have been achieved. This isn't about giving trade secrets away, but rather about allowing consumers to feel like they can trust these machines and defend themselves if an error were to occur. Although it is not based in artificial intelligence, a good examples of what this might look like is CreditKarma, which

allows people to have a better understanding of their credit score—a system that used to be hidden just like algorithms are today. This enables consumers to have a better understanding of what's happening behind the scenes and debate the legitimacy of their results if they believe the system has failed. Consumers should no longer be dictated by the invisible policy rules that algorithms have become; instead, machines should have to explain themselves as a matter of public safety.

Audit Algorithms

Regardless of how well the issues facing AI systems are addressed, much of what happens behind the scenes will continue to be beyond the understanding of the general public, something Larry Page and Sergey Brin once warned about in their PhD dissertation, where they stated, "Most search engine development has gone on at companies with little publication of technical details. This causes search engine technology to remain largely a black art."[14] Their observation is truer now than ever before, in part, because companies want to protect their intellectual property (IP) and make it difficult for malicious actors to game the system—a noble and legitimate concern; but also because they want to make it difficult to critically assess the inner workings of their algorithms because it's better for business.

To the defense of these companies, much of what is done involves very high-level mathematics, which most people would not be able to understand. Nevertheless, there are ways to allow the people to feel safe, one of which is to establish an ecosystem of people who the public can trust to audit the algorithms and evaluate potential concerns with the safety of the public in mind. Cathy O'Neil, founder of ORCAA, has begun spearheading this initiative independently, but we need more people involved, and to ensure they have the leverage to do their job appropriately, they need legislative support.[15] Without powerful allies, we run the risk that audits remain internal, which all too often leads to less than thorough inspection. Whether this happens because of insider bias or because the person responsible for auditing wants to avoid a hard conversation with whomever is in charge doesn't matter, these people should not be put in this position in the first place. Instead, audits should be performed by independent parties supported by those with power to ensure legitimate results.

1 These are the current feedback options given at the end of a Facebook Messenger phone call.

The problem with these options is that four of the five skew positive, which leads to a positive skew in Facebook's data, reinforcing their work whether it is deserved or not—a rare public-facing proof of insider bias.

The question that must then be asked is: *If these are the options given in a public facing survey, what questions were asked behind closed doors that led to their approval?*

POOR   FAIR   GOOD   VERY GOOD   EXCELLENT

To ensure objectivity, standards should also be created to define how auditors are assigned to companies. For example, audits should not be able to be done by the same company each time, to avoid the potential of regulatory capture, a process by which powerful organizations use their influence to manipulate the agencies that are supposed to control them. Having a diversity of organizations would not only ensure we negate regulatory capture, but would also allow these organizations to create a safe space to trade knowledge between each other and form an independent regulatory body that is more agile than the large, bureaucratic governing bodies that preside over most nations today. Audits should be required once companies reach a specified threshold of data collection and processing as a way to proactively monitor the system and promote honest behavior. Defining this threshold will be important to protect small companies that may otherwise struggle to innovate.

All Systems Go

Ultimately, our fear should not reside in the fact that we might create machines that can think for themselves or want to annihilate the human race. While these are realistic potential outcomes, it is much more likely that we create systems that are intelligent enough to be sold to consumers in a way that allows them to remain unquestioned, which, I believe, should be our real concern. If the complexity of life is reduced to logical relationships or substituted with proxy data, we risk severe damage to the fabric of human life as we know it. Poor search results may not kill anyone, but there is plenty to be concerned about if machines are given the right to make decisions they aren't prepared for and cannot be scrutinized.

However, if built with ethics, these systems have the potential to complement and amplify life in ways we can't even imagine yet. As the European Commission has stated in Legislative Act 157 of the General Data Protection Regulation (GDPR): "Research results obtained through registries [global databases] provide solid, high-quality knowledge which can provide the basis for the formulation and implementation of knowledge-based policy, improve the quality of life for a number of people, and improve the efficiency of social services."[16] However, to make this a reality we must ensure consumers can trust these systems, which means taking the necessary steps to earn their trust.

CHAPTER 17

Define Privacy

Most people around the world do not understand the impact that data and privacy abuses can have on their lives. The idea is incredibly intangible and abstract, but that should not be a reason to avoid ensuring everyone a basic level of personal security. Helping people understand and control their data, and installing safety measures to protect unknowing individuals, should be considered a basic human right.

17

Privacy is a fundamental human right that allows us to be our true selves. It allows us to be weirdos without shame, to have dissenting opinions without consequence, and, ultimately, to be free. This is why many nations have strict laws concerning privacy. In spite of this common understanding, privacy in the information age is one of the least understood and poorly defined topics because it spans a vast array of issues, taking shape in many different forms, which makes it incredibly difficult to (1) identify and (2) discuss. However, I'd like to try to resolve this ambiguity in one paragraph:

In the United States, it is a federal offense to open someone's mail. This is considered a criminal breach of privacy that could land someone in prison for up to five years.[1] Metaphorically speaking, each piece of data—whether photo, video, text, or something else—can be thought of as parcel of mail. However, unlike opening our mail in real life, internet companies can legally open every piece of mail that gets delivered through their system, without legal consequence. Moreover, they can make copies of it as well. What they're doing would be comparable to someone opening our mail, copying it at Kinkos, then storing it in a filing cabinet with our name on it and sharing it with anyone willing to pay for it. Want to open that filing cabinet or delete some of the copies? Too bad. Our mail is currently considered their property and we have almost no control over how it gets used.

In 1981 the Council of Europe made the first attempt to resolve these issues by establishing the "Convention for the Protection of Individual with Regard to Automatic Processing of Personal Data." In it, the council laid out principles that respected privacy as a "fundamental value" and gave individuals control over personal data. Since then, two dominant models have emerged, which can generally be understood by looking at the differences between regulation in the United States and Europe. On one side, we have the EU's recent General Data Protection Regulation (GDPR), which outlines the ways in which data should be processed and stored. On the other side, we have the flexible regulatory approach of the United States, which allows corporations and other private organizations to define privacy best practices and considers these companies to be the owners of the data.[2]

In general, scholars have noted the dangers of both models. The European model risks stifling innovation through regulation. The US model risks individual safety by putting power in the hands of for-profit organizations that have little or no financial incentive to do anything unless the public demands it. Many have argued that the US model is a fundamental threat to democracy without external incentives to push innovation, and looking at how people in the tech community feel about privacy reveals that there is some validity to this argument.[2]

Although it may not have been well known outside the tech world, Steve Jobs was a low-tech parent who limited his kids' use of technology.[3] Mark Zuckerberg spent $30 million to buy his house and the four houses surrounding it to maintain his privacy.[4] Elon Musk and Jeff Bezos are spending billions in an attempt to colonize Mars and get away from Earth. And many others in the tech world, both famous and not, are intentionally working to limit the time they spend online.[5] It's

clear that people in the tech world understand something about the value of privacy that the general public does not. This doesn't, however, mean that privacy is something citizens do not desire.

Findings by Pew Research Center show that 90 percent of adults believe it is important that they have control over what information is collected about them, 93 percent believe it's important they can control who has access to this information, and 86 percent have taken steps to remove or mask their digital footprints.[6][7] Similar numbers were discovered about Europeans in doteveryone's 2018 Digital Attitudes report.[8] Despite these numbers, 59 percent still feel like it is impossible to remain anonymous online, 68 percent believe current laws do not do enough to protect their privacy, and only 6 percent are "very confident" that government agencies can keep them secure.[9][10]

Giving up privacy is something the population has been forced into, due to the monopolies that exist in the tech world. Because of this, privacy has become something only the rich can afford, which is regrettable. If companies would spend half the money they spend trying to increase engagement on figuring out the best way to respect people's privacy, the problem would already be solved. But because that's not what's best for the bottom line, they continue operating without addressing the issue. Moving forward it will be important that these issues are considered by all parties involved. As Barack Obama noted in his administration's summary of concerns about privacy on the internet: "One thing should be clear, even though we live in a world in which we share personal information more freely than in the past, we must reject the conclusion that privacy is an outmoded value. It has been at the heart of our democracy from its inception, and we need it now more than ever."[11]

Implement Trust Networks

One of the first steps that needs to be taken to protect consumers is to make it easier for both consumers and third parties to access data. This may sound counterintuitive, but it is only because most people have yet to realize that demanding stricter controls over data would only reinforce Big Tech's data monopolies. Future regulation should, instead, enhance the free flow of data between platforms by working to create secure, seamless data exchanges that give consumers control of their data and enable startups to fight back. Doing this will be no small task but is something researchers such as Alex "Sandy" Pentland, creator and director of the MIT Media Lab, have been working on for quite some time now. In his book, *Social Physics: How Social Networks Can Make Us Smarter*, Pentland suggests that "a successful data-driven society must be able to guarantee that our data will not be abused—and perhaps especially that government will not abuse the power conferred by access to such fine-grain data," which is why he believes companies should implement what he is calling "trust networks."[12]

Trust networks, as Pentland describes them, keep track of user permissions for each piece of personal data, abiding by a legal contract that defines what can and can't be done with the data, as well as what happens in case of a violation of the terms. This would put people in control of their data and guarantee companies are held responsible for their use of the data they are given. While this may sound radical, it's not so different from what occurred less than a decade ago when consumers demanded protection from spam emails, which resulted in the CAN-SPAM Act—legislation that defined the rules for commercial email, gave recipients the right to have individuals and companies stop emailing them, and spelled out tough penalties for violations.[13] Giving people this level of control over their data will undoubtedly be the future of data collection as well, at which point Pentland believes, "Sharing personal data could become as safe and secure as transferring money between banks."

Today anyone can download their data from most platforms but it is valueless to anyone besides the platform it was downloaded from, which makes attempting to move from one platform to another like traveling internationally and hoping to use our local currency. Unfortunately, unlike money, there are no exchange operations for data on the internet at this point in history, meaning if consumers want to switch they have to start from scratch, which is difficult. For example, if someone wanted to move from Facebook to a new, more consumer-focused platform, they would have to download their data from Facebook, inspect what's there piece-by-piece, then place it in all the appropriate locations on the new platform, if those locations even exist. Nobody wants to do that. This is exactly why consumers don't try new platforms, the process is like moving to a new home and leaving everything in the old house, only to light it on fire and watch it burn to the ground on the way out.

To change this, consumers need a system that allows them to download their data from anywhere and seamlessly transfer it wherever they want with as little friction as possible. This would not only make it easier for consumers to leave one platform for another but would simultaneously force monopoly platform owners to innovate or risk being overtaken by new platforms working to protect consumers. The key to making this centralized exchange would be to ensure that it is created as a public utility that has world-class security and is legally incapable of using the data to create platforms or otherwise influence the global market in any way. Exchanges like this will be necessary if we want to reintroduce competition into the market and regain consumer trust.

Simplify Terms of Service Agreements

Another major problem with US privacy practices is how hard it is to understand terms and service agreements. Currently, people are forced to read long documents full of legal language and technical jargon if they hope to understand what they're agreeing to. One study actually demonstrated that it would take approximately 201 hours (nearly ten days) per year for the average person to read every privacy policy they encounter on an annual basis. The researchers estimated that the value of this lost time would amount to nearly $781 billion per year, which is unacceptable.[14] This puts consumers in a position where they're forced to opt-in without truly understanding what they're getting into. This could change if companies were required to implement what Gilad Rosner of the IoT Privacy Forum calls "usable privacy," a strategy that works to acquire people's consent as they interact with the system instead of demanding consent for everything up front.

However, while this strategy would make understanding privacy easier, there are also a couple concerns that need to be addressed. First, creating more steps along the way will induce friction to the experience, which may push consumers to pursue other products that offer a more frictionless experience. This can be negated with legislation that makes usable privacy required but wouldn't resolve the second, less obvious problem, which is that giving consumers greater control has actually been proven to provide them with a false sense of security, making it more likely that they'll give away more sensitive information.[15] Despite noble intentions, this means giving people more control may actually backfire, putting consumers in a position that is more dangerous than before.[16][17]

Understanding both sides of this dilemma makes it clear that there are no simple answers. While I do believe that consumers deserve tools that enable them to understand and control their data, increased safety will ultimately come down to consumer education, as the Centre for Information Policy Leadership has stated in their 2018 report, "Regulating for Results: Strategies and Priorities for Leadership and Engagement": "Unless individuals understand the importance of data protection, and can relate it to their own lives, it will never be fully effective."[18] Even more important is that companies hire people dedicated to making consumer privacy a priority, which will enable the necessary changes to happen.

Install Data Protection Officers (DPOs)

Tasked with overseeing and critiquing projects that involve data processing and collection, data protection officers (DPOs) play a crucial role in mitigating privacy risks, developing an internal culture that embraces privacy by design, and avoiding adverse publicity by building trust and assisting with legal compliance—all factors that save companies money. These officers, and the teams they hire, should be highly knowledgeable about the ways in which data is used and how to protect consumer privacy. They should have good connections in the industry and be capable of acting with diplomacy when representing companies worldwide. And they should be capable of running regular privacy impact assessments (PIAs) to proactively prevent mistakes by regulating internally.[19] Whether hired by the company or brought in as outside consultants, DPOs should be the companies' North Star when it comes to defining best practices in privacy and security.

In studies where DPOs were interviewed, two keys were identified as factors that make their job possible. The first being that they are capable of remaining independent, even if they work within the companies, and second, that they have the ability to report directly to a

1 Giving people more control over their data and expecting it to make the internet safer is like putting a nutrition label on a Snickers bar and expecting it to be less fattening.

board member or senior executive.[20] By remaining contractually independent, DPOs are allowed to be part of the company but not so close as to develop insider bias. This, combined with the ability to report directly to higher-level staff members, enables officers to remain neutral while still having authority.

The next step is to make sure they have the funding necessary to operate without restraint or fear of regulatory capture, as noted in Article 52(4) of the GDPR: "Each Member State shall ensure that each supervisory authority is provided with the human, technical and financial resources, premises and infrastructure necessary for the effective performance of its tasks and exercise of its powers..." It has not, however, been determined where exactly this funding should come from, which can make it hard to secure, as proven by a 2016 study that found the average budget for data protection authorities (DPAs) in Europe was around €8 per business, or €0.41 per individual citizen.[21] This is something that could be reversed by collecting an annual fee of just €20 from all organizations that collect and process data, according to the Centre for Information Policy Leadership.[22] To bolster the authority of DPAs, legislative measures should be implemented to increase financial support for their work.

Incentivize Change

Creating trustworthy and secure data-sharing experiences will be one of the biggest challenges our world will face in the coming decades. Thus, discovering creative ways to make these issues a priority for all stakeholders should be considered essential for businesses and policymakers alike; a change that may require financial incentives. For example, tax breaks could be created for companies that allocate large amounts of their budget to improving their systems. They could be given to companies that decide to supply regular training or workshops for their staff to help make privacy and security a priority in the company culture. They could be given to companies that hire professional hackers to find loopholes in their systems before attacks occur. In this sense, such incentives would not be so different than tax breaks given to businesses that implement eco-friendly practices.

The idea of tax breaks may sound outrageous, but incentives such as these would represent a more proactive solution than the way things are handled now. While it may feel good to read a headline stating, "Google fined a record $5 billion by the EU for Android antitrust violations," we must keep in mind that fines like this only represent a small fraction of such companies' revenue.[23] Combine this with the fact that most cases take several years or decades to conclude, and that percentage only gets smaller. With this for consideration, the idea of tax breaks can be approached from a different perspective, which is that they are not about rewarding previously negligent behavior but about increasing public safety in a way that is in the best interest of everyone involved. Maintaining our current system, which allows companies to string out court cases while they continue their malpractices is just as, if not more, dangerous than having no laws at all.

CHAPTER 18

Update Consumer Protection Standards

The internet is moving faster than most people are capable of truly understanding or keeping up with. Until data and technology literacy around the world has reached a level of general understanding, creating some bumper lanes to keep things in line should be considered a matter of public safety.

Currently, the web is largely unregulated, which has resulted in some of the greatest technologies ever seen. Never has it been so easy to start a business. Never before have we been capable of collaborating with teammates in remote locations around the globe in real time. And never before has it been so frictionless to maintain communication with loved ones around the globe. However, in recognizing the awe-inspiring innovation that has occurred in the last 20 to 30 years we also have to recognize how dangerous things have become. Malevolent actors are capable of manipulating audio and video to the point where the truth is unrecognizable, internet companies are strategically investing in ways to addict people to their services, and malicious bots threaten the safety of everyone involved. Worse yet, this is all being done with little or no public awareness. Consequently, it's time we have a conversation about what needs to be fixed.

Define Criminal Bot Behavior

For better or worse, bots play a large role in the internet. We need them for many reasons but there are also a lot of bots we would be better off without, many of which have reached levels of specialized intelligence beyond human capabilities. Consider that we already have systems that outperform humans on a wide variety of tasks, including image recognition and synthesis, speech recognition, proprioception, and air combat simulation, among other things.[1][2][3] So as a large group of AI experts warn in their report, "The Malicious Use of Artificial Intelligence: Forecasting, Prevention, and Mitigation": "We are thus at a critical moment in the co-evolution of AI and cybersecurity and should proactively prepare for the next wave of attacks."[4]

While the threats these systems pose can be breathtaking, any concerns about their potential for destruction are not unlike concerns people have had with previous technologies. Airplanes, which would have been nothing more than unidentified flying objects to previous generations, caused incredible fear at first, but now most of us don't think twice about jumping on a plane to travel. The same can be said of biochemical and nuclear weapons, which have the potential to cause overwhelming destruction but have been tamed by our security and defense practices. Each of these technologies represented radical ideas that caused fears that have since been dampened through focused, collaborative efforts designed to discover ways to negate their dangers, and the same will likely be true of artificial intelligence.

Two of the most comprehensive reports on the possibilities of automated threats are "The Malicious Use of Artificial Intelligence: Forecasting, Prevention and Mitigation," referred to previously, and "Artificial Intelligence and National Security," written by a group of Harvard researchers well versed in the field. These researchers propose several changes, including mandating that policymakers work with technical researchers to understand the problems; creating a system that allows researchers and practitioners to raise a warning flag without tarnishing their reputation by investigating malicious approaches or enabling bad actors to prematurely discover these flaws and leverage the vulnerabilities; and bringing in participants from other industries to spur conversations that may not have resulted if left only to professionals with deep expertise.[5] These reports list many other potential solutions and should be taken into consideration as we work to define protective standards.

Protect Against Manipulative Design Patterns

Another danger we need to address is how to eliminate manipulative design patterns, which first requires that we identify them. To start, have you ever noticed how every time you open or refresh Facebook, Twitter, or Instagram there's something new at the top of the screen? Have you ever had something disappear after accidentally refreshing, only to struggle to find it again? Or maybe you've noticed how easy it is to get lost in an endless stream of YouTube content? These experiences are intentionally designed to spur the same dopamine releases we receive pulling the lever on a slot machine, which creates what the industry calls a "sticky" experience and drives fear of missing out (FOMO) so we spend more time on their platforms. Even if we're not looking at ads, even if we don't read a single post from anyone, even if we're just scrolling mindlessly, we're technically engaging with the platform, and companies can use these numbers to sell their ads at a higher price. So, they continue to use manipulative design patterns despite the fact that they're doing serious damage to public well-being across the world.

Continuing, regulation should also be created in regards to coercive patterns. For example, have you ever noticed how Google's ads have evolved to appear almost as if they were not paid for? The same can be seen in FB, Instagram, Twitter, and nearly every other platform that has decided to create "in-stream" ads. Advertisements should be required to be blatantly obvious, otherwise consumers' perceptions will silently be shaped by whoever has the most money. But the coercive patterns do not end there. For more, we can look at the way terms and service agreements are often designed. Notice how there's almost never a "I don't agree, but I'd still like to use the service" button? Even if consumers have to pay for a service, they should have the option to not agree with everything contained in terms and service agreements. The idea of these patterns is to get people to do things that are beneficial to the bottom line, even if that might not be in their best interest.

In order to negate these dangers, design patterns such as the ones listed above, as well as others, should be formally outlawed as a matter of consumer protection for several reasons. First, being put into legislation would guarantee these protections for consumers, automatically making the internet more trustworthy. At the same time, appropriate legislation would prevent businesses interested in deceiving consumers from undermining those that decide to willfully make the positive changes consumers deserve. Those who have adapted or are adapting would face no consequences, but with laws like this in place companies would be forced to implement these changes or risk punishment, which would be invaluable to consumers. To determine what should and should not be outlawed, we can look to people like Tristan Harris and company at the Humane Center for Computing; Ame Elliot

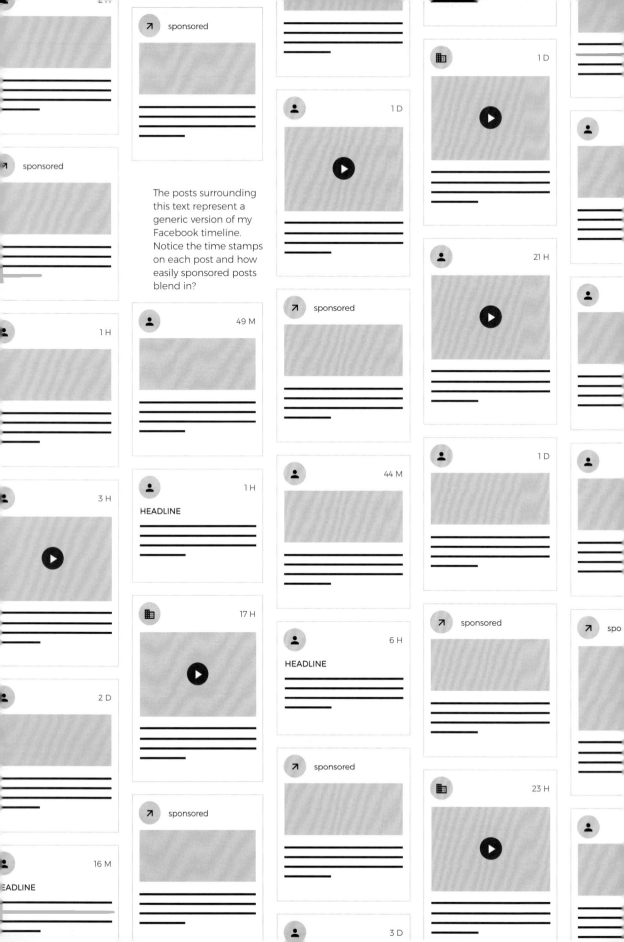

The posts surrounding this text represent a generic version of my Facebook timeline. Notice the time stamps on each post and how easily sponsored posts blend in?

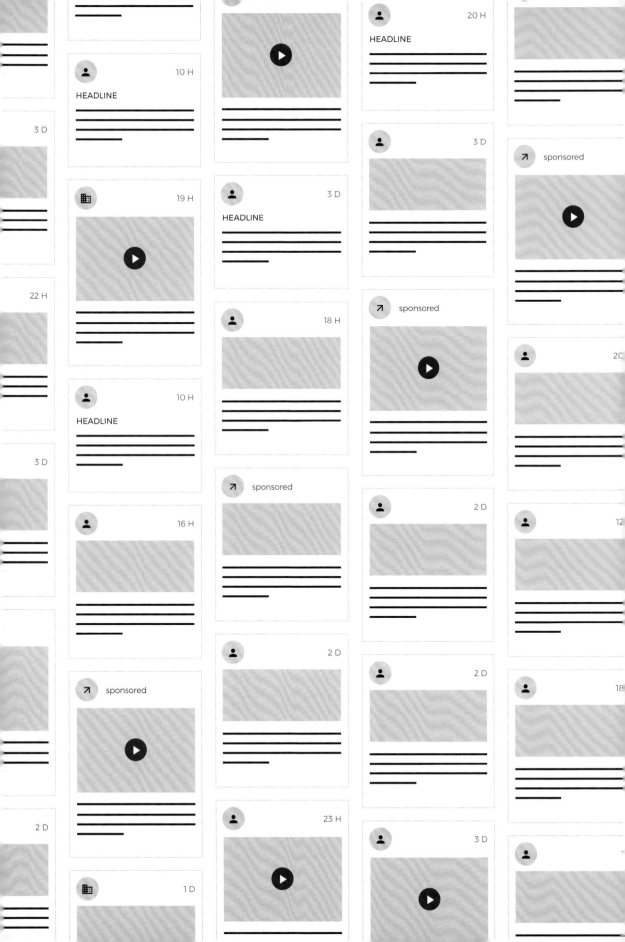

and team at Simply Secure; and Nir Eyal, who has learned a lot about what not to do through his experiences; along with many others who have been working to informally define these dangers for years.

Enforcement should also include regular audits, just like audits for malicious algorithms, which should be done by expert designers who are capable of, and professionally authorized to, audit design systems. These people and their organizations should become a contact point for concerned consumers to reach out to if they're uncomfortable or recognize something that may have been missed. Participation from the general public will be critical, as no one service or group of services alone will be able to keep up with the needs of the internet.

Require Consumer Education

Regardless of how well new laws protect consumers, there is still a fundamental problem underlying everything going on with the internet, which is that most consumers don't fully understand what's happening due to the speed at which innovation is occurring—a risk to everyone involved. It is a risk to consumers who can be easily manipulated, and it is a risk to companies that may end up getting regulated too tightly due to misunderstandings within a consumer population they've abused. Only through an informed and empowered public will we be capable of maintaining flexible regulation, as Tim Wu made clear in his book *The Master Switch: The Rise and Fall of Information Empires*:

> *While the government is critical as a bulwark and countervailing weight to private power, it is not ultimately with government that the best hope for robust Separations regime ultimately lies. As important as any regulatory principle (whether enforced by carrot or stick) is an underlying ethical principle, one demanded by the public and enforced by norms of conduct within industry.*[6]

A key fact to recognize is that the same abusive cycle we've seen with modern internet companies has been seen before. In fact, it happened as recently as the late 1990s and early 2000s with the US tobacco industry. After mountains of evidence was brought to the attention of the public, including public service campaigns unveiling the secrets of Big Tobacco, consumers began to demand change. While tobacco remains legal, as much of the behavior of internet companies should, one of the biggest changes that resulted in world-altering impact was legislation that forced Big Tobacco to educate the public about the dangers of their products. Since this legislation was put in place, the United States has seen cigarette usage decrease by more than half.[7]

As with tobacco usage, there is now a great deal of evidence that points to the damage being imposed on the general population by Big Tech. To decrease such damage, Big Tech companies should be forced to educate consumers about the dangers of the internet. This education should not only include Public Service Announcements (PSAs), but also require companies to use their behavioral data to

Visual Evolution of Google Ad Design

† Where do the ads begin and where do they end?

monitor for signs of addictive behavior and provide assistance to the individuals they identify. For example, Netflix uses its behavioral data to monitor people and reaches out if they have been watching an excessive amount of shows. One man they reached out to stated that having a company look out for his well-being, "Made [him] feel better just knowing that someone, even a stranger working at a customer support agency, cared about [his] mental health."[8] This sentiment validates what JANA Partners and CALSTRS, major investors in Apple, have told Apple Inc, which is that, "Paying special attention to the health and development of the next generation is both good business and the right thing to do."[9] To ensure companies are invested in this, legislation should require them to fund research focused on discovering healthy engagement levels to provide them with objective safety benchmarks.

If the public remains unaware of what's happening and doesn't have tools to help them understand what dangers are present, they will remain in a position that is ripe for manipulation. This has never been truer than in today's climate, which has been intentionally created to favor corporations over consumers. Ensuring the growing presence of an educated public that is capable of identifying dangers and holding corporations accountable could perhaps be considered one of today's most important goals. Forcing companies to pay for that education and provide consumers with the tools necessary to act are two simple steps that could very well change the world.

Power to the People

While it would be ideal if companies embraced their responsibilities, this, as we've seen, doesn't always happen. And when consumers are incapable of pointing out flaws, this makes it easy for companies to represent themselves as something they may not be, which goes against everything we'd hope for when giving these organizations flexibility in the first place. Instead of leveraging this freedom for public good, companies have corrupted this gift by using it to abuse consumers in a way many are incapable of understanding. What's happening across the internet today may not be war crimes, but it is only a matter of time before the situation gets worse; this is not a matter of if, but when, and we have a responsibility to prepare.

The difficulty in deciding upon what should and should not be regulated is that as soon as any legislation is passed, the industry will transform—companies will move somewhere that doesn't enforce the laws, find ways around the regulation, or otherwise reinvent themselves as we've seen many times before. This makes lawmakers, especially those who fail to understand the fundamentals of what's happening, hesitant to create and pass legislation. But until we have informed consumers who are capable of recognizing danger and defending themselves, consumer protection standards must be updated to meet the demands of our modern world. While it will not be easy, failing to at least create some bumper lanes for the industry would, at best, be negligent on the part of those in charge of ensuring public safety.

CHAPTER 19

Improve Education

Right now we're witnessing a clash of the old and new worlds as we transition through this new Industrial Revolution. While many would love to turn back the clock, that's not going to happen. The best way to prepare citizens for this future is with an equally revolutionary change in education. These changes must be made if we want to ensure a level of stability in people's lives.

In September 2013, two Oxford researchers, Carl Benedikt Frey and Michael A. Osborne, published a paper titled "The Future of Employment," in which they evaluated the likelihood of different professions being eliminated due to automation over the next 10 to 20 years. They concluded that 47 percent of US jobs are at high risk.[1] On a global scale, the McKinsey Institute predicted in 2017 that between 400 and 800 million jobs (about the population of the United States and Europe combined) are at risk of being automated by 2030.[2] Others have predicted less, around 14 percent, according to the OECD in 2018, because they believe jobs will be created as well.[3] The exact number of jobs that will be lost to automation is something we won't know until it happens, but what we can be certain of is that change is coming.

The antidote, according to many experts around the world, is education, which, it is said, should enable students to become economically independent; understand the world outside of their own bubbles; engage in a civil manner with their fellow citizens; and, ultimately, help the world grow in a meaningful way. This is a lot to ask of any educational institution, but is even more concerning considering we've entered an era where our world is perpetually changing. Because of this, revamping our education system to meet the needs of the modern world will be, as Ginni Rommetty, CEO of IBM, has stated, "the issue of our time."[4] The challenge educators face today resides in discovering the best way to educate current students for a rapidly evolving marketplace while simultaneously retraining tens of millions of workers who will more than likely be displaced by automation.[5] To achieve these goals, our education systems must work to broaden students' minds and reduce standardization in favor of a more creative education that allows students to explore the human side of life and prepares them for the modern world.

Reduce Standardization

Although the original purpose of schools was to enlighten and educate students, school quickly became standardized during the first Industrial Revolution to meet the needs of the market. This led many schools to focus on high grades as their measure of success. Unfortunately, the skills required to get good grades are not necessarily the same as the skills required to understand subjects, and because of this, decades of standardized education has led to a world in which schools are systematically standardizing entire populations. While this may be good for the economy, it's not necessarily what's best for creating engaged students and citizens. In fact, standardization has led to the United States having one of the highest dropout rates of any developed nation despite the fact that this issue was formally recognized by the US Department of Education under the Reagan administration in a 1983 report titled, "A Nation At Risk," in which the department stated:[6]

> The educational foundations of our society are presently being eroded by a rising tide of mediocrity that threatens our very future as a Nation and a people...If an unfriendly foreign power had attempted to

impose on America the mediocre educational performance that exists today, we might well have viewed it as an act of war. As it stands, we have allowed this to happen to ourselves.[7]

It has been estimated that if the dropout rate in the US were cut in half, it would save the country nearly $90 billion per year based on the money that would be saved from social programs and earned from having a higher number of employed citizens.[8] If someone were running a business that was losing this many customers and creating such a financial deficit, most business owners would try to figure out a solution. Instead, the United States has chosen to continue forward without flinching. The result of standardizing such mediocrity has led to the US being placed at number 24 in both reading and science, and number 40 in mathematics out of nearly 70 developed nations surveyed in 2015 by the Programme for International Student Assessment (PISA).[9] Worse yet, the country was delisted from Bloomberg's top ten most innovative countries worldwide as of early 2018.[10] Yet when we look at what's going on in policy discussions nationwide it appears as if nobody is concerned. This a dangerous mindset but it wouldn't be the first time we've seen it, the same thing happened in Japan, which was a world leader in innovation throughout the late 1980s and early 1990s.

After finding what the country considered to be the perfect education system, Japan, too, standardized its process. Unfortunately, this resulted in a system that standardized the innovation out of its people—a point that has been written about by several Asian studies scholars, including Akiko Hashimoto, associate professor of Sociology and Asian Studies at the University of Pittsburgh, who stated, "The Ministry of Education's approach—encouraging creativity through rules and regulations will never succeed, for there is an inherent contradiction in trying to 'mold' young people into freethinking individuals."[11] As troubling as this is to read, it shouldn't be shocking when a system designed to achieve a specific goal achieves that goal. Unfortunately, this is what will happen to any nation that systematically pursues specialization for too long. Eventually, the population will be filled with individuals who, like specialized forms of artificial intelligence, are relatively incapable of handling problems outside of their specialization. And that's exactly what has happened in the United States.

However, things don't have to continue this way. An example of how to maintain measurable progress while not focusing solely on measurements and standardization is what is happening in Finland, a nation that sat at the top of PISA's 2015 rankings, reaching number 4 in reading, number 5 in science and number 13 in mathematics, respectively.[12] Despite what many may think, the country actively works to avoid standardization. In fact, there is only one standardized test in Finland, and it comes at the end of high school.

While Finland does offer a standard curriculum, it is very broad and balanced, including everything from sciences and mathematics to art, languages, and humanities. The thing that differentiates Finland is that while a standardized menu of courses is offered, the final decision about what will be taught is ultimately determined by principals, on

Average scores of 15-year-old students on the PISA science literacy scale, by education system: 2015

1.	*Singapore*	36.	*Lithuania*
2.	Japan	37.	*Croatia*
3.	Estonia	38.	*Buenos Aires (Argentina)*
4.	*Chinese Taipei*	39.	Iceland
5.	Finland	40.	Israel
6.	*Macau (China)*	41.	*Malta*
7.	Canada	42.	Slovak Republic
8.	*Vietnam*	43.	Greece
9.	*Hong Kong (China)*	44.	Chile
10.	*B-S-J-G (China)*	45.	*Bulgaria*
11.	Korea, Republic of	46.	*United Arab Emirates*
12.	New Zealand	47.	*Uruguay*
13.	Slovenia	48.	*Romania*
14.	Australia	49.	*Cyprus*
15.	United Kingdom	50.	*Moldova, Republic of*
16.	Germany	51.	*Albania*
17.	Netherlands	52.	Turkey
18.	Switzerland	53.	*Trinidad and Tobago*
19.	Ireland	54.	*Thailand*
20.	Belgium	55.	*Costa Rica*
21.	Denmark	56.	*Qatar*
22.	Poland	57.	*Colombia*
23.	Portugal	58.	Mexico
24.	Norway	59.	*Montenegro, Republic of*
25.	**United States**	60.	*Georgia*
26.	Austria	61.	*Jordan*
27.	France	62.	*Indonesia*
28.	Sweden	63.	*Brazil*
29.	Czech Republic	64.	*Peru*
30.	Spain	65.	*Lebanon*
31.	Latvia	66.	*Tunisia*
32.	*Russian Federation*	67.	*Macedonia, Republic of*
33.	Luxembourg	68.	*Kosovo*
34.	Italy	69.	*Algeria*
35.	Hungary	70.	*Dominican Republic*

NOTE: Education systems are ordered by 2015 average score. The OECD average is the average of the national averages of the OECD member countries, with each country weighted equally. Scores are reported on a scale from 0 to 1,000. Italics indicate non-OECD countries and education systems. B-S-J-G (China) refers to the four PISA participating China provinces: Beijing, Shanghai, Jiangsu, and Guangdong. Although Argentina, Malaysia, and Kazakhstan participated in PISA 2015, technical problems with their samples prevent results from being discussed in this report. The standard errors of the estimates are shown in table R1 available at http://nces.ed.gov/surveys/pisa/PISA2015/index.asp.

SOURCE: Organization for Economic Cooperation and Development (OECD), Program for International Student Assessment (PISA), 2015.

Average scores of 15-year-old students on the PISA reading literacy scale, by education system: 2015

1.	*Singapore*	36.	Luxembourg
2.	*Hong Kong (China)*	37.	Israel
3.	Canada	38.	*Buenos Aires (Argentina)*
4.	Finland	39.	*Lithuania*
5.	Ireland	40.	Hungary
6.	Estonia	41.	Greece
7.	Korea, Republic of	42.	Chile
8.	Japan	43.	Slovak Republic
9.	Norway	44.	*Malta*
10.	New Zealand	45.	*Cyprus*
11.	Germany	46.	*Uruguay*
12.	*Macau (China)*	47.	*Romania*
13.	Poland	48.	*United Arab Emirates*
14.	Slovenia	49.	*Bulgaria*
15.	Netherlands	50.	Turkey
16.	Australia	51.	*Costa Rica*
17.	Sweden	52.	*Trinidad and Tobago*
18.	Denmark	53.	*Montenegro, Republic of*
19.	France	54.	*Colombia*
20.	Belgium	55.	Mexico
21.	Portugal	56.	*Moldova, Republic of*
22.	United Kingdom	57.	*Thailand*
23.	*Chinese Taipei*	58.	*Jordan*
24.	**United States**	59.	*Brazil*
25.	Spain	60.	*Albania*
26.	Russian Federation	61.	*Qatar*
27.	*B-S-J-G (China)*	62.	*Georgia*
28.	Switzerland	63.	*Peru*
29.	Latvia	64.	*Indonesia*
30.	Czech Republic	65.	*Tunisia*
31.	Croatia	66.	*Dominican Republic*
32.	*Vietnam*	67.	*Macedonia, Republic of*
33.	Austria	68.	*Algeria*
34.	Italy	69.	*Kosovo*
35.	Iceland	70.	*Lebanon*

NOTE: Education systems are ordered by 2015 average score. The OECD average is the average of the national averages of the OECD member countries, with each country weighted equally. Scores are reported on a scale from 0 to 1,000. Italics indicate non-OECD countries and education systems. B-S-J-G (China) refers to the four PISA participating China provinces: Beijing, Shanghai, Jiangsu, and Guangdong. Although Argentina, Malaysia, and Kazakhstan participated in PISA 2015, technical problems with their samples prevent results from being discussed in this report. The standard errors of the estimates are shown in table R1 available at http://nces.ed.gov/surveys/pisa/PISA2015/index.asp.

SOURCE: Organization for Economic Cooperation and Development (OECD), Program for International Student Assessment (PISA), 2015.

Average scores of 15-year-old students on the PISA mathematics literacy scale, by education system: 2015

1.	*Singapore*	36.	*Lithuania*
2.	*Hong Kong (China)*	37.	Hungary
3.	*Macau (China)*	38.	Slovak Republic
4.	*Chinese Taipei*	39.	Israel
5.	Japan	**40.**	**United States**
6.	*B-S-J-G (China)*	41.	*Croatia*
7.	Korea, Republic of	42.	*Buenos Aires (Argentina)*
8.	Switzerland	43.	Greece
9.	Estonia	44.	*Romania*
10.	Canada	45.	*Bulgaria*
11.	Netherlands	46.	*Cyprus 437*
12.	Denmark	47.	*United Arab Emirates*
13.	Finland	48.	Chile
14.	Slovenia	49.	Turkey
15.	Belgium	50.	*Moldova, Republic of*
16.	Germany	51.	*Uruguay*
17.	Poland	52.	*Montenegro, Republic of*
18.	Ireland	53.	*Trinidad and Tobago*
19.	Norway	54.	*Thailand*
20.	Austria	55.	*Albania*
21.	New Zealand	56.	Mexico
22.	*Vietnam*	57.	*Georgia*
23.	*Russian Federation*	58.	*Qatar*
24.	Sweden	59.	*Costa Rica*
25.	Australia	60.	*Lebanon*
26.	France	61.	*Colombia*
27.	United Kingdom	62.	*Peru*
28.	Czech Republic	63.	*Indonesia*
29.	Portugal	64.	*Jordan*
30.	Italy	65.	*Brazil*
31.	Iceland	66.	*Macedonia, Republic of*
32.	Spain	67.	*Tunisia*
33.	Luxembourg	68.	*Kosovo*
34.	Latvia	69.	*Algeria*
35.	*Malta*	70.	*Dominican Republic*

NOTE: Education systems are ordered by 2015 average score. The OECD average is the average of the national averages of the OECD member countries, with each country weighted equally. Scores are reported on a scale from 0 to 1,000. Italics indicate non-OECD countries and education systems. B-S-J-G (China) refers to the four PISA participating China provinces: Beijing, Shanghai, Jiangsu, and Guangdong. Although Argentina, Malaysia, and Kazakhstan participated in PISA 2015, technical problems with their samples prevent results from being discussed in this report. The standard errors of the estimates are shown in table R1 available at http://nces.ed.gov/surveys/pisa/PISA2015/index.asp.

SOURCE: Organization for Economic Cooperation and Development (OECD), Program for International Student Assessment (PISA), 2015.

Bloomberg 2018 Innovation Index

1.	South Korea		26.	Malaysia
2.	Sweden		27.	Hungary
3.	Singapore		28.	Czech Republic
4.	Germany		29.	Spain
5.	Switzerland		30.	Portugal
6.	Japan		31.	Greece
7.	Finland		32.	Luxembourg
8.	Denmark		33.	Turkey
9.	France		34.	Lithuania
10.	Israel		35.	Romania
11.	**United States**		36.	Estonia
12.	Austria		37.	Hong Kong
13.	Ireland		38.	Slovakia
14.	Belgium		39.	Malta
15.	Norway		40.	Latvia
16.	Netherlands		41.	Bulgaria
17.	United Kingdom		42.	Croatia
18.	Australia		43.	Tunisia
19.	China		44.	Serbia
20.	Italy		45.	Thailand
21.	Poland		46.	Ukraine
22.	Canada		47.	Cyprus
23.	New Zealand		48.	South Africa
24.	Iceland		49.	Iran
25.	Russia		50.	Morocco

Notes: The 2018 ranking process began with more than 200 economies. Each was scored on a 0-100 scale based on seven equally weighted categories. Nations that didn't report data for at least six categories were eliminated, trimming the list to 80. Bloomberg released the top 50 and category scores within this cohort.

1. R&D intensity: Research and development expenditure, as % GDP 2. **Manufacturing value-added:** MVA, as % GDP and per capita ($PPP) 3. **Productivity:** GDP and GNI per employed person age 15+ and 3Y improvement 4. **High-tech density:** Number of domestically dommiciled high-tech public companies—such as aerospace and defense, biotechnology hardware, software, semiconductors, internet software and services, and renewable energy companies—as % domestic publicly listed companies and as a share of world's total public high-tech companies 5. **Tertiary efficiency:** Total enrollment in tertiary education, regardless of age, as % the post-secondary cohort; share of labor force with advanced level of education; annual new science and engineering graduates as % total tertiary graduates and as % the labor force 6. **Research concentration:** Professionals, including postgraduate PhD students, engaged in R&D per million population 7. **Patent activity:** Resident patent filings, total patent grants and patent in force, per million population; filings per $100 billion GDP and total grants by country as a share of world total.

All metric are equally weighted. Metrics consisting of multiple factors were rescaled for countries void of some but not all data points. Most recent data available used. Of the more than 200 economies evaluated, 80 had data available for at least six of the seven factors and were ranked; The top 50 and the metric ranks among them are displayed.

Sources: Bloomberg, International Labour Organization, International Monetary Fund, World Bank, Organization for Economic Co-operation and Development, World Intellectual Property Organization

a local level. The country also invests heavily in teacher development, which not only makes them better at their jobs but also helps raise the cultural status of being a good teacher. Finland also encourages schools to collaborate, rather than compete, by sharing resources, ideas, and expertise among both school districts and community stakeholders.[13] In his book *Creative Schools*, comprehensive assessment of the current state of education, Sir Ken Robinson, "Finnish schools do not do these things in addition to achieving high standards—they achieve high standards precisely because they do these things."[14]

A similar education system exists in South Korea, which led the world in Bloomberg's 2018 innovation index and was also ranked highly by PISA: seven in reading, eleven in science, and seven in mathematics.[15][16] Instead of standardizing, South Korean education is focused on finding a balance by teaching students, "key competencies that creative and integrative learners should acquire, such as self-management competency, knowledge-information processing skills, creative thinking skills, aesthetic-emotional competency, communication skills, and civic competency." By teaching students things machines are not good at, the country encourages "creative and integrative learners" who are prepared to work in a world where our traditional economic burdens have become automated.[17]

The speed at which specific jobs are automated will have a lot to do with how easy the skill is to teach to a machine versus how much financial value will be returned by creating the system. Because of this, many of the "soft" human skills that cannot be standardized will become exceedingly valuable in the new economy. And we don't have to rely on anecdotes to validate this; these skills have been proven to make the most successful workers. Even Google, with all of its technological prowess, discovered this to be the truth when it ran an experiment called Project Oxygen in 2008 to analyze what kind of people are most valuable to its company. Despite the logical hypothesis that those with the greatest STEM expertise would be the most valuable, the top performing employees turned out to be those who were capable of: being a good coach; communicating and listening well; possessing insights into others (including others' different values and points of view); having empathy toward and being supportive of one's colleagues; being a good critical thinker and problem solver; and being able to make connections across complex ideas. STEM expertise came in dead last.[18][19]

A similar study was also run by IBM in 2008, in which the company spoke with 1,500 leaders in 80 countries to discover what characteristics organization leaders need most in their staff. The two priorities they discovered were adaptability to change and creativity in generating new ideas. They also found that these qualities were often lacking in many otherwise highly qualified graduates.[20] The truth in these findings is that few, if any, of the abilities possessed by the best employees are capable of being standardized. These skills, which are also difficult to teach a machine, will only become more relevant as many "hard" skills quickly become automated. As we attempt to reform education to meet the needs of this new world, the fundamental goals of creating high-quality citizens who are capable of being economically independent and culturally adept should remain, but the way we achieve these goals has to change.

Increase Technology and Data Literacy

We've all heard the phrase "Knowledge is power." Throughout history many leaders and organizations have used knowledge, in the form of literacy, to establish and maintain power over large populations. Only after becoming literate could people form their own opinions about life and how it should be lived. But despite the fact that most people believe literacy is now universal, illiteracy still exists. In fact, in 2014 the United States Central Intelligence Agency (CIA) reported that there were still more than 750 million illiterate adults around the world, nearly 10 percent of the population. Of these people, more than 75 percent were from South Asia or Sub-Saharan Africa and more than 66 percent were women, demonstrating how the residual effects of systematic inequality can span centuries.[21]

Literacy not only allows people to communicate with others, but also allows them to participate in the economy, which for centuries now has been dominated by the English language because of the power and opportunity that resides in English-speaking markets like those of the United States and Europe. Today, the internet and mass adoption of computers has spawned a new form of literacy. And just like English before it, this new literacy and the languages contained within it will only continue to become more dominant over time due to market demands. So, despite the argument that has been put forth thus far, our education systems should not solely be focused on soft skills. Students will need hard skills that allow them to critically analyze the world, and none will be more important than technology and data literacy in the future we're headed towards.

Whether students decide to work in science, engineering, technology, and math (STEM) fields or not, teaching them the language of the future will give them flexibility in life and allow them to find stability in the shifting tides of our modern economy—similar to how reading and writing have enabled previous generations. As Martha Lane Fox, founder of doteveryone and member of the House of Lords has said: "Building our understanding can help give us resilience in a time of technological change. With greater understanding—as individuals and as a society—we will be better able to harness the opportunities technologies present."[22]

Jump-Start Education

There have always been debates about what should be taught in school. In ancient Rome, education was based on the seven liberal arts or sciences: grammar, the formal structures of language; rhetoric, composition, and presentation of argument; dialectic, formal logic; arithmetic; geometry; music; and astronomy. During the Renaissance, the curriculum expanded as some schools began to teach music, dancing, drawing, and sports. Later, in the 18th century, the curriculum began to focus on history, geography, mathematics, and foreign languages. Then, after the onset of the Industrial Revolution, standardization was emphasized to meet the needs of the market. With this, humans began to be trained as machines, prepared for work from a young age, and now, two centuries later, we are still following the same equation.[23]

Debates about what should and should not be included in the curriculum will always exist and are necessary to ensure our education systems meet the needs of our current reality. But no matter what is debated, there's no debating that it's time for a change. As Andreas Schleicher, special advisor on education policy to the secretary-general at the OECD has made clear, "The world economy no longer pays you for what you know; Google knows everything. The world economy pays you for what you can do with what you know."[24] The challenge will be meeting these new needs while ensuring students are capable of flourishing in a rapidly evolving, data driven economy. Making this change will be no small task, but with the right foundation it can be achieved.

The Robot Ate My Job

The following is an alphabetized list of jobs and their likelihood of being automated by 2032, as forecasted by Frey and Osborne[25]

Accountants and Auditors	94%	Ambulance Drivers and Attendants, Except Emergency Medical Technicians	25%
Actors	37%		
Actuaries	21%		
Adhesive Bonding Machine Operators and Tenders	95%	Amusement and Recreation Attendants	72%
Administrative Law Judges, Adjudicators, and Hearing Officers	64%	Animal Breeders	95%
		Animal Control Workers	21%
Administrative Services Managers	73%	Animal Scientists	6.1%
Adult Basic and Secondary Education and Literacy Teachers and Instructors	19%	Animal Trainers	10%
		Anthropologists and Archeologists	0.77%
		Appraisers and Assessors of Real Estate	90%
Advertising and Promotions Managers	3.9%	Arbitrators, Mediators, and Conciliators	6%
Advertising Sales Agents	54%		
Aerospace Engineering and Operations Technicians	48%	Architects, Except Landscape and Naval	1.8%
Aerospace Engineers	1.7%	Architectural and Civil Drafters	52%
Agents and Business Managers of Artists, Performers, and Athletes	24%	Architectural and Engineering Managers	1.7%
		Archivists	76%
Agricultural and Food Science Technicians	97%	Art Directors	2.3%
Agricultural Engineers	49%	Astronomers	4.1%
Agricultural Inspectors	94%	Athletes and Sports Competitors	28%
Air Traffic Controllers	11%	Athletic Trainers	0.71%
Aircraft Cargo Handling Supervisors	6.6%	Atmospheric and Space Scientists	67%
Aircraft Mechanics and Service Technicians	71%	Audio and Video Equipment Technicians	55%
Aircraft Structure, Surfaces, Rigging, and Systems Assemblers	79%	Audio-Visual and Multimedia Collections Specialists	39%
		Audiologists	0.33%
Airfield Operations Specialists	71%	Automotive and Watercraft Service Attendants	83%
Airline Pilots, Copilots, and Flight Engineers	18%		

Automotive Body and Related Repairers	91%
Automotive Glass Installers and Repairers	55%
Automotive Service Technicians and Mechanics	59%
Avionics Technicians	70%
Baggage Porters and Bellhops	83%
Bailiffs	36%
Bakers	89%
Barbers	80%
Bartenders	77%
Bicycle Repairers	94%
Bill and Account Collectors	95%
Billing and Posting Clerks	96%
Biochemists and Biophysicists	2.7%
Biological Scientists, All Other	1.5%
Biological Technicians	30%
Biomedical Engineers	3.7%
Boilermakers	68%
Bookkeeping, Accounting, and Auditing Clerks	98%
Brickmasons and Blockmasons	82%
Bridge and Lock Tenders	97%
Broadcast News Analysts	6.7%
Broadcast Technicians	74%
Brokerage Clerks	98%
Budget Analysts	94%
Bus and Truck Mechanics and Diesel Engine Specialists	73%
Bus Drivers, School or Special Client	89%
Bus Drivers, Transit and Intercity	67%
Business Operations Specialists, All Other	23%
Butchers and Meat Cutters	93%
Buyers and Purchasing Agents, Farm Products	87%
Cabinetmakers and Bench Carpenters	92%
Camera and Photographic Equipment Repairers	97%
Camera Operators, Television, Video, and Motion Picture	60%
Captains, Mates, and Pilots of Water Vessels	27%
Cardiovascular Technologists and Technicians	23%
Career/Technical Education Teachers, Middle School	26%

Career/Technical Education Teachers, Secondary School	0.88%
Cargo and Freight Agents	99%
Carpenters	72%
Carpet Installers	87%
Cartographers and Photogrammetrists	88%
Cashiers	97%
Cement Masons and Concrete Finishers	94%
Chefs and Head Cooks	10%
Chemical Engineers	1.7%
Chemical Equipment Operators and Tenders	76%
Chemical Plant and System Operators	85%
Chemical Technicians	57%
Chemists	10%
Chief Executives	1.5%
Child, Family, and School Social Workers	2.8%
Childcare Workers	8.4%
Chiropractors	2.7%
Choreographers	0.4%
Civil Engineering Technicians	75%
Civil Engineers	1.9%
Claims Adjusters, Examiners, and Investigators	98%
Cleaners of Vehicles and Equipment	37%
Cleaning, Washing, and Metal Pickling Equipment Operators and Tenders	81%
Clergy	0.81%
Clinical, Counseling, and School Psychologists	0.81%
Coaches and Scouts	1.3%
Coating, Painting, and Spraying Machine Setters, Operators, and Tenders	91%
Coil Winders, Tapers, and Finishers	73%
Coin, Vending, and Amusement Machine Servicers and Repairers	94%
Combined Food Preparation and Serving Workers, Including Fast Food	92%
Commercial and Industrial Designers	3.7%
Commercial Divers	18%
Commercial Pilots	55%
Compensation and Benefits Managers	96%

Compensation, Benefits, and Job Analysis Specialists	47%
Compliance Officers	8%
Computer and Information Research Scientists	1.5%
Computer and Information Systems Managers	3.5%
Computer Hardware Engineers	22%
Computer Numerically Controlled Machine Tool Programmers, Metal and Plastic	36.6%
Computer Occupations, All Other	22%
Computer Operators	78%
Computer Programmers	48%
Computer Support Specialists	65%
Computer Systems Analysts	0.65%
Computer-Controlled Machine Tool Operators, Metal and Plastic	86%
Computer, Automated Teller, and Office Machine Repairers	74%
Concierges	21%
Conservation Scientists	1.6%
Construction and Building Inspectors	63%
Construction and Related Workers, All Other	71%
Construction Laborers	88%
Construction Managers	7.1%
Continuous Mining Machine Operators	54%
Control and Valve Installers and Repairers, Except Mechanical Door	63%
Conveyor Operators and Tenders	93%
Cooks, Fast Food	81%
Cooks, Institution and Cafeteria	83%
Cooks, Private Household	30%
Cooks, Restaurant	96%
Cooks, Short Order	94%
Cooling and Freezing Equipment Operators and Tenders	93%
Correctional Officers and Jailers	60%
Correspondence Clerks	86%
Cost Estimators	57%
Costume Attendants	61%
Counter and Rental Clerks	97%
Counter Attendants, Cafeteria, Food Concession, and Coffee Shop	96%
Couriers and Messengers	94%
Court Reporters	50%
Court, Municipal, and License Clerks	46%
Craft Artists	3.5%
Crane and Tower Operators	90%
Credit Analysts	98%
Credit Authorizers, Checkers, and Clerks	97%
Credit Counselors	4%
Crossing Guards	49%
Crushing, Grinding, and Polishing Machine Setters, Operators, and Tenders	97%
Curators	0.68%
Customer Service Representatives	55%
Cutters and Trimmers, Hand	64%
Cutting and Slicing Machine Setters, Operators, and Tenders	86%
Cutting, Punching, and Press Machine Setters, Operators, and Tenders, Metal and Plastic	78%
Dancers	13%
Data Entry Keyers	99%
Database Administrators	3%
Demonstrators and Product Promoters	51%
Dental Assistants	51%
Dental Hygienists	68%
Dental Laboratory Technicians	97%
Dentists, General	0.44%
Derrick Operators, Oil and Gas	80%
Desktop Publishers	16%
Detectives and Criminal Investigators	34%
Diagnostic Medical Sonographers	35%
Dietetic Technicians	13%
Dietitians and Nutritionists	0.39%
Dining Room and Cafeteria Attendants and Bartender Helpers	91%
Directors, Religious Activities and Education	91%
Dishwashers	77%
Dispatchers, Except Police, Fire, and Ambulance	96%
Door-to-Door Sales Workers, News and Street Vendors, and Related Workers	94%
Dredge Operators	92%
Drilling and Boring Machine Tool Setters, Operators, and Tenders, Metal and Plastic	94%
Driver/Sales Workers	98%
Drywall and Ceiling Tile Installers	79%

Earth Drillers, Except Oil and Gas	85%	Engine and Other Machine Assemblers	82%
Economists	43%	Engineering Technicians, Except Drafters, All Other	24%
Editors	5.5%	Engineers, All Other	1.4%
Education Administrators, Elementary and Secondary School	0.46%	Environmental Engineering Technicians	25%
Education Administrators, Postsecondary	1%	Environmental Engineers	1.8%
Education Administrators, Preschool and Childcare Center/Program	1.5%	Environmental Science and Protection Technicians, Including Health	77%
Educational, Guidance, School, and Vocational Counselors	0.85%	Environmental Scientists and Specialists, Including Health	3.3%
Electric Motor, Power Tool, and Related Repairers	76%	Epidemiologists	3.3%
Electrical and Electronic Equipment Assemblers	95%	Etchers and Engravers	98%
Electrical and Electronics Drafters	81%	Excavating and Loading Machine and Dragline Operators	94%
Electrical and Electronics Engineering Technicians	84%	Executive Secretaries and Executive Administrative Assistants	86%
Electrical and Electronics Installers and Repairers, Transportation Equipment	91%	Explosives Workers, Ordnance Handling Experts, and Blasters	48%
Electrical and Electronics Repairers, Commercial and Industrial Equipment	41%	Extruding and Drawing Machine Setters, Operators, and Tenders, Metal and Plastic	91%
Electrical and Electronics Repairers, Powerhouse, Substation, and Relay	38%	Extruding and Forming Machine Setters, Operators, and Tenders, Synthetic and Glass Fibers	88%
Electrical Engineers	10%	Extruding, Forming, Pressing, and Compacting Machine Setters, Operators, and Tenders	93%
Electrical Power-Line Installers and Repairers	9.7%		
Electricians	15%		
Electro-Mechanical Technicians	81%	Fabric and Apparel Patternmakers	0.49%
Electromechanical Equipment Assemblers	97%	Fabric Menders, Except Garment	96%
		Fallers	76%
Electronic Equipment Installers and Repairers, Motor Vehicles	61%	Farm and Home Management Advisors	0.75%
Electronic Home Entertainment Equipment Installers and Repairers	65%	Farm Equipment Mechanics and Service Technicians	75%
		Farm Labor Contractors	97%
Electronics Engineers, Except Computer	2.5%	Farmers, Ranchers, and Other Agricultural Managers	4.7%
Elementary School Teachers, Except Special Education	0.44%	Fashion Designers	2.1%
		Fence Erectors	92%
Elevator Installers and Repairers	39%	Fiberglass Laminators and Fabricators	93%
Eligibility Interviewers, Government Programs	70%	File Clerks	97%
Embalmers	54%	Film and Video Editors	31%
Emergency Management Directors	0.3%	Financial Analysts	23%
Emergency Medical Technicians and Paramedics	4.9%	Financial Examiners	17%
		Financial Managers	6.9%
		Financial Specialists, All Other	33%

Fine Artists, Including Painters, Sculptors, and Illustrators	4.2%
Fire Inspectors and Investigators	48%
Firefighters	17%
First-Line Supervisors of Construction Trades and Extraction Workers	17%
First-Line Supervisors of Correctional Officers	2.5%
First-Line Supervisors of Farming, Fishing, and Forestry Workers	57%
First-Line Supervisors of Fire Fighting and Prevention Workers	0.36%
First-Line Supervisors of Food Preparation and Serving Workers	63%
First-Line Supervisors of Helpers, Laborers, and Material Movers, Hand	42%
First-Line Supervisors of Housekeeping and Janitorial Workers	94%
First-Line Supervisors of Landscaping, Lawn Service, and Groundskeeping Workers	57%
First-Line Supervisors of Mechanics, Installers, and Repairers	0.3%
First-Line Supervisors of Non-Retail Sales Workers	7.5%
First-Line Supervisors of Office and Administrative Support Workers	1.4%
First-Line Supervisors of Personal Service Workers	7.6%
First-Line Supervisors of Police and Detectives	0.44%
First-Line Supervisors of Production and Operating Workers	1.6%
First-Line Supervisors of Retail Sales Workers	28%
First-Line Supervisors of Transportation and Material-Moving Machine and Vehicle Operators	2.9%
Fish and Game Wardens	8%
Fishers and Related Fishing Workers	83%
Fitness Trainers and Aerobics Instructors	8.5%
Flight Attendants	35%
Floor Layers, Except Carpet, Wood, and Hard Tiles	79%
Floor Sanders and Finishers	87%
Floral Designers	4.7%
Food and Tobacco Roasting, Baking, and Drying Machine Operators and Tenders	91%
Food Batchmakers	70%
Food Cooking Machine Operators and Tenders	61%
Food Preparation Workers	87%
Food Scientists and Technologists	7.7%
Food Servers, Nonrestaurant	86%
Food Service Managers	8.3%
Forensic Science Technicians	0.95%
Forest and Conservation Technicians	42%
Forest and Conservation Workers	87%
Forest Fire Inspectors and Prevention Specialists	4.8%
Foresters	0.81%
Forging Machine Setters, Operators, and Tenders, Metal and Plastic	93%
Foundry Mold and Coremakers	67%
Funeral Attendants	37%
Funeral Service Managers, Directors, Morticians, and Undertakers	20%
Furnace, Kiln, Oven, Drier, and Kettle Operators and Tenders	37%
Furniture Finishers	87%
Gaming and Sports Book Writers and Runners	91%
Gaming Cage Workers	39%
Gaming Change Persons and Booth Cashiers	83%
Gaming Dealers	96%
Gaming Managers	9.1%
Gaming Supervisors	28%
Gaming Surveillance Officers and Gaming Investigators	95%
Gas Compressor and Gas Pumping Station Operators	91%
Gas Plant Operators	78%
General and Operations Managers	16%
Geographers	25%
Geological and Petroleum Technicians	91%
Geoscientists, Except Hydrologists and Geographers	63%
Glaziers	73%
Graders and Sorters, Agricultural Products	41%
Graphic Designers	8.2%
Grinding and Polishing Workers, Hand	97%

Laborers and Freight, Stock, and Material Movers, Hand	85%
Landscape Architects	4.5%
Landscaping and Groundskeeping Workers	95%
Lathe and Turning Machine Tool Setters, Operators, and Tenders, Metal and Plastic	84%
Laundry and Dry-Cleaning Workers	71%
Lawyers	3.5%
Layout Workers, Metal and Plastic	84%
Legal Secretaries	98%
Librarians	65%
Library Assistants, Clerical	95%
Library Technicians	99%
Licensed Practical and Licensed Vocational Nurses	5.8%
Life, Physical, and Social Science Technicians, All Other	61%
Lifeguards, Ski Patrol, and Other Recreational Protective Service Workers	67%
Light Truck or Delivery Services Drivers	69%
Loading Machine Operators, Underground Mining	50%
Loan Interviewers and Clerks	92%
Loan Officers	98%
Locker Room, Coatroom, and Dressing Room Attendants	43%
Locksmiths and Safe Repairers	77%
Locomotive Engineers	96%
Locomotive Firers	93%
Lodging Managers	0.39%
Log Graders and Scalers	97%
Logging Equipment Operators	79%
Logisticians	1.2%
Machine Feeders and Offbearers	93%
Machinists	65%
Maids and Housekeeping Cleaners	69%
Mail Clerks and Mail Machine Operators, Except Postal Service	94%
Maintenance and Repair Workers, General	64%
Maintenance Workers, Machinery	86%
Makeup Artists, Theatrical and Performance	1%
Management Analysts	13%
Managers, All Other	25%
Manicurists and Pedicurists	95%

Manufactured Building and Mobile Home Installers	18%
Marine Engineers and Naval Architects	1%
Market Research Analysts and Marketing Specialists	61%
Marketing Managers	1.4%
Marriage and Family Therapists	1.4%
Massage Therapists	54%
Materials Engineers	2.1%
Materials Scientists	2.1%
Mathematical Technicians	99%
Mathematicians	4.7%
Meat, Poultry, and Fish Cutters and Trimmers	94%
Mechanical Door Repairers	91%
Mechanical Drafters	68%
Mechanical Engineering Technicians	38%
Mechanical Engineers	1.1%
Medical and Clinical Laboratory Technicians	47%
Medical and Clinical Laboratory Technologists	90%
Medical and Health Services Managers	0.73%
Medical Appliance Technicians	45%
Medical Assistants	30%
Medical Equipment Preparers	78%
Medical Equipment Repairers	27%
Medical Records and Health Information Technicians	91%
Medical Scientists, Except Epidemiologists	0.45%
Medical Secretaries	81%
Medical Transcriptionists	89%
Meeting, Convention, and Event Planners	3.7%
Mental Health and Substance Abuse Social Workers	0.31%
Mental Health Counselors	0.48%
Merchandise Displayers and Window Trimmers	48%
Metal-Refining Furnace Operators and Tenders	88%
Meter Readers, Utilities	85%
Microbiologists	1.2%
Middle School Teachers, Except Special and Career/Technical Education	17%

Milling and Planing Machine Setters, Operators, and Tenders, Metal and Plastic	98%
Millwrights	59%
Mine Cutting and Channeling Machine Operators	59%
Mine Shuttle Car Operators	37%
Mining and Geological Engineers, Including Mining Safety Engineers	14%
Miscellaneous Agricultural Workers	87%
Mixing and Blending Machine Setters, Operators, and Tenders	83%
Mobile Heavy Equipment Mechanics, Except Engines	40%
Model Makers, Metal and Plastic	93%
Model Makers, Wood	96%
Models	98%
Molders, Shapers, and Casters, Except Metal and Plastic	90%
Molding, Coremaking, and Casting Machine Setters, Operators, and Tenders, Metal and Plastic	95%
Motion Picture Projectionists	97%
Motorboat Mechanics and Service Technicians	66%
Motorboat Operators	62%
Motorcycle Mechanics	79%
Multimedia Artists and Animators	1.5%
Multiple Machine Tool Setters, Operators, and Tenders, Metal and Plastic	91%
Museum Technicians and Conservators	59%
Music Directors and Composers	1.5%
Musical Instrument Repairers and Tuners	91%
Musicians and Singers	7.4%
Natural Sciences Managers	1.8%
Network and Computer Systems Administrators	3%
New Accounts Clerks	99%
Nonfarm Animal Caretakers	82%
Nuclear Engineers	7%
Nuclear Medicine Technologists	13%
Nuclear Power Reactor Operators	95%
Nuclear Technicians	85%

Occupational Health and Safety Specialists	17%
Occupational Health and Safety Technicians	25%
Occupational Therapists	0.35%
Occupational Therapy Aides	27%
Occupational Therapy Assistants	2.8%
Office Clerks, General	96%
Office Machine Operators, Except Computer	92%
Operating Engineers and Other Construction Equipment Operators	95%
Operations Research Analysts	3.5%
Ophthalmic Laboratory Technicians	97%
Opticians, Dispensing	71%
Optometrists	14%
Oral and Maxillofacial Surgeons	0.36%
Order Clerks	98%
Orthodontists	2.3%
Orthotists and Prosthetists	0.35%
Outdoor Power Equipment and Other Small Engine Mechanics	93%
Packaging and Filling Machine Operators and Tenders	98%
Packers and Packagers, Hand	38%
Painters, Construction and Maintenance	75%
Painters, Transportation Equipment	69%
Painting, Coating, and Decorating Workers	92%
Paper Goods Machine Setters, Operators, and Tenders	67%
Paperhangers	87%
Paralegals and Legal Assistants	94%
Parking Enforcement Workers	84%
Parking Lot Attendants	87%
Parts Salespersons	98%
Patternmakers, Metal and Plastic	90%
Patternmakers, Wood	91%
Paving, Surfacing, and Tamping Equipment Operators	83%
Payroll and Timekeeping Clerks	97%
Personal Care Aides	74%
Personal Financial Advisors	58%
Pest Control Workers	66%
Pesticide Handlers, Sprayers, and Applicators, Vegetation	97%
Petroleum Engineers	16%

Petroleum Pump System Operators, Refinery Operators, and Gaugers	71%
Pharmacists	1.2%
Pharmacy Aides	72%
Pharmacy Technicians	92%
Photographers	2.1%
Photographic Process Workers and Processing Machine Operators	99%
Physical Scientists, All Other	43%
Physical Therapist Aides	61%
Physical Therapist Assistants	1.8%
Physical Therapists	2.1%
Physician Assistants	14%
Physicians and Surgeons	0.42%
Physicists	10%
Pile-Driver Operators	82%
Pipelayers	62%
Plant and System Operators, All Other	86%
Plasterers and Stucco Masons	84%
Plating and Coating Machine Setters, Operators, and Tenders, Metal and Plastic	92%
Plumbers, Pipefitters, and Steamfitters	35%
Podiatrists	0.46%
Police and Sheriff's Patrol Officers	9.8%
Police, Fire, and Ambulance Dispatchers	49%
Political Scientists	3.9%
Postal Service Clerks	95%
Postal Service Mail Carriers	68%
Postal Service Mail Sorters, Processors, and Processing Machine Operators	79%
Postmasters and Mail Superintendents	75%
Postsecondary Teachers	3.2%
Pourers and Casters, Metal	87%
Power Distributors and Dispatchers	64%
Power Plant Operators	85%
Prepress Technicians and Workers	97%
Preschool Teachers, Except Special Education	0.74%
Pressers, Textile, Garment, and Related Materials	81%
Print Binding and Finishing Workers	95%
Printing Press Operators	83%
Private Detectives and Investigators	31%
Probation Officers and Correctional Treatment Specialists	25%
Procurement Clerks	98%
Producers and Directors	2.2%
Production Workers, All Other	92%
Production, Planning, and Expediting Clerks	88%
Proofreaders and Copy Markers	84%
Property, Real Estate, and Community Association Managers	81%
Prosthodontists	5.5%
Psychiatric Aides	47%
Psychiatric Technicians	4.3%
Psychologists, All Other	0.43%
Public Address System and Other Announcers	72%
Public Relations and Fundraising Managers	1.5%
Public Relations Specialists	18%
Pump Operators, Except Wellhead Pumpers	90%
Purchasing Agents, Except Wholesale, Retail, and Farm Products	77%
Purchasing Managers	3%
Radiation Therapists	34%
Radio and Television Announcers	10%
Radio Operators	98%
Radio, Cellular, and Tower Equipment Installers and Repairs	93%
Radiologic Technologists and Technicians	23%
Rail Car Repairers	88%
Rail Yard Engineers, Dinkey Operators, and Hostlers	91%
Rail-Track Laying and Maintenance Equipment Operators	89%
Railroad Brake, Signal, and Switch Operators	83%
Railroad Conductors and Yardmasters	83%
Real Estate Brokers	97%
Real Estate Sales Agents	86%
Receptionists and Information Clerks	96%
Recreation Workers	0.61%
Recreational Therapists	0.28%
Recreational Vehicle Service Technicians	59%
Refractory Materials Repairers, Except Brickmasons	82%

Refuse and Recyclable Material Collectors	93%
Registered Nurses	0.9%
Rehabilitation Counselors	0.94%
Reinforcing Iron and Rebar Workers	90%
Reporters and Correspondents	11%
Reservation and Transportation Ticket Agents and Travel Clerks	61%
Residential Advisors	6.4%
Respiratory Therapists	6.6%
Respiratory Therapy Technicians	10%
Retail Salespersons	92%
Riggers	89%
Rock Splitters, Quarry	96%
Rolling Machine Setters, Operators, and Tenders, Metal and Plastic	83%
Roof Bolters, Mining	49%
Roofers	90%
Rotary Drill Operators, Oil and Gas	53%
Roustabouts, Oil and Gas	68%
Sailors and Marine Oilers	83%
Sales Engineers	0.41%
Sales Managers	1.3%
Sales Representatives, Wholesale and Manufacturing, Except Technical and Scientific Products	85%
Sales Representatives, Wholesale and Manufacturing, Technical and Scientific Products	25%
Sawing Machine Setters, Operators, and Tenders, Wood	86%
Secondary School Teachers, Except Special and Career/Technical Education	0.78%
Secretaries and Administrative Assistants, Except Legal, Medical, and Executive	96%
Securities, Commodities, and Financial Services Sales Agents	1.6%
Security and Fire Alarm Systems Installers	82%
Security Guards	84%
Segmental Pavers	83%
Self-Enrichment Education Teachers	13%
Semiconductor Processors	88%
Separating, Filtering, Clarifying, Precipitating, and Still Machine Setters, Operators, and Tenders	88%

Septic Tank Servicers and Sewer Pipe Cleaners	83%
Service Unit Operators, Oil, Gas, and Mining	93%
Set and Exhibit Designers	0.55%
Sewers, Hand	99%
Sewing Machine Operators	89%
Shampooers	79%
Sheet Metal Workers	82%
Ship Engineers	4.1%
Shipping, Receiving, and Traffic Clerks	98%
Shoe and Leather Workers and Repairers	52%
Shoe Machine Operators and Tenders	97%
Signal and Track Switch Repairers	90%
Skincare Specialists	29%
Slaughterers and Meat Packers	60%
Slot Supervisors	54%
Social and Community Service Managers	0.67%
Social and Human Service Assistants	13%
Social Science Research Assistants	65%
Social Scientists and Related Workers, All Other	4%
Sociologists	5.9%
Software Developers, Applications	4.2%
Software Developers, Systems Software	13%
Soil and Plant Scientists	2.1%
Sound Engineering Technicians	13%
Special Education Teachers, Middle School	1.6%
Special Education Teachers, Secondary School	0.77%
Speech-Language Pathologists	0.64%
Stationary Engineers and Boiler Operators	89%
Statistical Assistants	66%
Statisticians	22%
Stock Clerks and Order Fillers	64%
Stonemasons	89%
Structural Iron and Steel Workers	83%
Structural Metal Fabricators and Fitters	41%
Substance Abuse and Behavioral Disorder Counselors	3.3%
Subway and Streetcar Operators	86%
Surgical Technologists	34%
Survey Researchers	23%

Surveying and Mapping Technicians	96%
Surveyors	38%
Switchboard Operators, Including Answering Service	96%
Tailors, Dressmakers, and Custom Sewers	84%
Tank Car, Truck, and Ship Loaders	72%
Tapers	62%
Tax Examiners and Collectors, and Revenue Agents	93%
Tax Preparers	99%
Taxi Drivers and Chauffeurs	89%
Teacher Assistants	56%
Teachers and Instructors, All Other	0.95%
Team Assemblers	97%
Technical Writers	89%
Telecommunications Equipment Installers and Repairers, Except Line Installers	36%
Telecommunications Line Installers and Repairers	49%
Telemarketers	99%
Telephone Operators	97%
Tellers	98%
Terrazzo Workers and Finishers	88%
Textile Bleaching and Dyeing Machine Operators and Tenders	97%
Textile Cutting Machine Setters, Operators, and Tenders	95%
Textile Knitting and Weaving Machine Setters, Operators, and Tenders	73%
Textile Winding, Twisting, and Drawing Out Machine Setters, Operators, and Tenders	96%
Tile and Marble Setters	75%
Timing Device Assemblers and Adjusters	98%
Tire Builders	94%
Tire Repairers and Changers	70%
Title Examiners, Abstractors, and Searchers	99%
Tool and Die Makers	84%
Tool Grinders, Filers, and Sharpeners	88%
Tour Guides and Escorts	91%
Traffic Technicians	90%
Training and Development Managers	0.63%

Training and Development Specialists	1.4%
Transit and Railroad Police	57%
Transportation Attendants, Except Flight Attendants	75%
Transportation Inspectors	90%
Transportation, Storage, and Distribution Managers	59%
Travel Agents	9.9%
Travel Guides	5.7%
Tree Trimmers and Pruners	77%
Umpires, Referees, and Other Sports Officials	98%
Upholsterers	39%
Urban and Regional Planners	13%
Ushers, Lobby Attendants, and Ticket Takers	96%
Veterinarians	3.8%
Veterinary Assistants and Laboratory Animal Caretakers	86%
Veterinary Technologists and Technicians	2.9%
Waiters and Waitresses	94%
Watch Repairers	99%
Water and Wastewater Treatment Plant and System Operators	61%
Weighers, Measurers, Checkers, and Samplers, Recordkeeping	95%
Welders, Cutters, Solderers, and Brazers	94%
Welding, Soldering, and Brazing Machine Setters, Operators, and Tenders	61%
Wellhead Pumpers	84%
Wholesale and Retail Buyers, Except Farm Products	29%
Woodworking Machine Setters, Operators, and Tenders, Except Sawing	97%
Word Processors and Typists	81%
Writers and Authors	3.8%
Zoologists and Wildlife Biologists	30%

CHAPTER 20

Fight for Representation

Every system has to find a balance that allows it to thrive. That balance is currently non-existent in the Valley. If we hope to turn this around, we need to push for equality by design, not by quota. If this initiative is denied, we must demand better.

Every complicated system has impactful foundational elements that are difficult to recognize until they've been removed because of their silent or invisible nature. But sometimes waiting until the element has disappeared is too late. An example of this is the story of gray wolves in Yellowstone National Park, which begins in the late 1800s when a group of biological scientists decided to explore Yellowstone. The purpose of the expedition was to document the details of Yellowstone's biological ecosystem and discover the best ways to create stability within the environment. What they discovered beyond a beautiful ecosystem of fauna and flora, was that the park was riddled by an abundance of gray wolves, which they assumed would have a negative impact on the park. To combat this, settlers were encouraged to eradicate all the wolves in the park, a task accomplished by the 1920s.[1]

Despite the concerns of these well-intentioned scientists and citizens, the impact of this eradication had the opposite effect on Yellowstone. After eradication of their primary predator, the Elk population exploded, which led to free-range grazing that wiped out many of the natural plant species, increased erosion, and fundamentally altered the biological ecosystem beyond recognition.[2] Scientists began to grow concerned, and in the 1990s decided to slowly reintroduce gray wolves in an attempt to resolve these problems, despite the wishes of local farmers and area residents. What happened shocked everyone. Instead of causing damage, the wolves began hunting elk again, which helped maintain a healthy population and protect against overgrazing, thereby allowing plant and animal species to repopulate previously devastated areas, ultimately resulting in the redirection of several streams and rivers within the park due to synergistic effects.[3]

This story provides a rare glimpse into the impact one variable can have on a complicated system, demonstrating that even the best intentions can lead to catastrophic damage when changes are made to a system that is not fully comprehensible. In spite of this, scientists and engineers at some of the top tech companies are attempting a similar experiment in which they are attempting to replicate, manipulate, and control the efficiencies of the most complicated system our world has to offer: life itself. The difficulty in understanding and recreating such a system is not in analyzing it and replicating what is immediately seen, but in appropriately translating those findings into a logical system that does not oversimplify complicated issues out of idealistic optimism, like what happened on a small scale in Yellowstone. Continuing forward, it will be critical that systems accurately reflect the natural diversity of life if they are to thrive.

Diversity of Sex and Gender

Today the outstanding majority of the tech workforce is male—between 70 and 80 percent, according to most studies, and as high as 88 percent in engineering roles, according to ComputerScience.org. Most peoples' first reaction to this disparity is to conclude that tech companies have unfair hiring practices, but this isn't entirely true. According to that same study by ComputerScience.org, 17.6 percent of students

graduating college with a computer science degree are women, which means that while the percentage of women in tech is low it is a relatively decent reflection of the talent available.[4] Upon further consideration, the real problem seems to be embedded in the cultural fabric of society, residing in the fact that women are treated differently due to unconscious biases many of us hold, women included.

In fact, a study conducted in 2014 asked 716 women why they left tech, and the overwhelming majority did not say "because math is hard," but that the work environment was no longer something they cared to participate in.[5] Another survey of more than 5,300 women that same year found the same thing.[6] Regardless of your opinion on the matter, this isn't a fluke or the result of biased survey practices. Behavioral studies have proven that investors are more likely to prefer a pitch by a man than a woman; that women who ask for a higher salary tend to be rated as more difficult to work with and less nice than men who do the same; and that both men and women are more likely to hire a male applicant than a woman.[10,11,12] Despite the fact that such behavior is often attributed to men, these findings cross gender boundaries, pointing to the fact that this is a societal problem embedded in the way we're socialized. Recognizing that we have biases that conflict with our values does not make anyone a bad person; it's human nature to have biases, but once such flaws are discovered, we should work to change them, and the best place to start is with children.

According to ComputerScience.org, not a single girl took the AP Computer Science test in Montana, Mississippi, or Wyoming in 2014.[10] This is the beginning of a funnel that now ends with just under 18 percent of computer science degrees being earned by women, which clearly demonstrates the truth behind our diversity dilemma: the fact that women are told from a young age that computers and mathematics are not for them. And then, when they try to get involved, they get harassed, forced to act like one of the boys, or are otherwise isolated, which puts them in a position where they're forced to either endure the abuse or walk away from a life they never had the opportunity to explore and learn to love.

To change this, we need little girls to grow up believing in their talents and wanting to work in tech. This means we need more female leaders they can look to as role models. And the best way to make this happen is not only by hiring women but by making a conscious effort to support them, from early childhood on. Pointing to poor hiring practices or the low number of women studying computer science are only superficial remediations to a much deeper problem, not unlike trying to apply a Band-Aid to a gunshot wound. Moving forward, it's important to recognize that all genders have valuable knowledge to bring to the workplace, to acknowledge our biases that may impact their ability to perform, and to start making changes.

Diversity of Race, Ethnicity, and Culture

According to recent research on diversity in Silicon Valley, women are not alone in their exclusion. From 2007 to 2015, it was discovered that

Women Still Get Paid Less Despite Higher Graduation Rate

Women have been graduating college at a higher rate than men since 2012 but their earnings remain well below mens.[11]

 Men's Annual Salary

Women's Annual Salary

$55,000

$50,000

$45,000

$40,000

$35,000

$30,000

$25,000

Years, 1960–2016

1960   1965   1970   1975   1980   1985   1990   1995   2000   2005   2010   2015

white people were more than twice as likely to become executives than their non-white counterparts, and, in fact, black and Hispanic representation actually declined.[12] Some claim a biological difference is driving this disparity, but such an argument is simply incoherent. An overview of history shows non-white cultures have been responsible for some of the greatest inventions of all time. The Sumerians were the first to establish language.[13] The formulation of numbers can be attributed to Arabs.[14] And it is the Egyptians we have to thank for the foundations of math.[15] None of what we're doing today would be possible without these fundamental discoveries, yet they are forgotten when the time comes for recognition.

Furthermore, in the United States nearly 33 percent of all patents granted and 25 percent of all Nobel Prizes awarded have been given to immigrants, despite the fact that they account for only 13 percent of the population.[16] Within Silicon Valley, 51 percent of startups worth more than a billion dollars have at least one immigrant founder, 71 percent of which had at least one immigrant in a key growth position.[17] Moreover, companies founded by first and second generation immigrants include: Apple, Google, Amazon, Facebook, Oracle, IBM, Uber, Airbnb, Yahoo, Intel, EMC, eBay, SpaceX, VMWare, AT&T, Tesla, NVIDIA, Qualcomm, Paypal, ADP, Reddit, SlackHQ, WeWork, Stripe, Cognizant, Intuit, 3M.[18] As Richard Florida, author of *The Flight of the Creative Class: The New Global Competition for Talent*, notes in his book:

> " *[The United States] doesn't have some intrinsic advantage in the production of creative people, new ideas, or startup companies. Its real advantage lies in its ability to attract these economic drivers from around the world. Of critical importance to the American success in this last century has been a tremendous influx of global talent.*[19]

Disregarding all of this, talent in Silicon Valley remains overwhelmingly white until there's a quota to be met, which leads to a culture in which even when hired, non-white individuals are forced to fit in. Sure, there are laws against such treatment, but just because these behaviors have become less evident doesn't mean they've disappeared. Instead, inequality has become much more insidious, hidden behind closed doors and driven into places that are difficult to trace, forcing these individuals to operate one way in public and another in private, silencing the things that made them unique in the first place. And by remaining silent we've allowed society to become psychologically entrapped in a war against "the other," despite the fact that there is no validity to such behavior. If left unaddressed, these issues will continue to be ingrained in the mental structures that socialize our children and perpetuate such conditions for generations to come.

Diversity of Knowledge and Ability

The issue of diversity does not, however, end with sex, race, or any other measure of socioeconomic status. It also must include equality of knowledge and ability. Lacking resources, being poor, having decreased mobility, or dealing with other involuntary hurdles are situations that no well-educated, able-bodied, upper-class child will grow up to understand. Nevertheless, the majority of those recruited for major tech firms come from affluent backgrounds, have Ivy League educations, and grew up in a bubble with little, if any, exposure to the diversity of life. No matter how well-intentioned these individuals may be, they will never fully understand the diversity of life the way those who live it every day do, which means products are destined to reflect the homogeneous group of individuals they were built by. When building systems to wrap around the globe and shape society, it is important that teams are comprised of people who understand the wider societal impact of what is being created. This is not only an ethical concern, but also backed by research, which proves that more diverse teams tend to generate more revenue.[20]

No Implementation Without Representation

While a considerable amount of research proves the value of diversity, there is also a much darker side to diversity: the fact that small, visible differences often result in great conflict that lasts lifetimes. In 2017, James Damore, a former Google engineer, became the most sensationalized unemployed person on the planet for suggesting that we might find explanations for why the industry suffers from such disparities by looking to the biological differences between populations. In his writing Damore made broad claims that avoided any connection to socialization. He continued by writing that in order to solve our problems we should "[d]e-emphasize empathy" because remaining "emotionally unengaged helps us better reason about the facts," a rationalization that reflects those made by soldiers fighting a war.[21]

The truth revealed by Damore's memo is that Silicon Valley has become so data-driven that it is incapable of seeing the flaws and biases of the data it is consuming, resulting in a cold, war-like environment that is now creating dangerous products without regard for the populations they're being built for, a practice that should not be acceptable. In 1775 the United States fought the American Revolution for the right to be represented in regards to taxation, as emphasized in the phrase all Americans are taught in history class, which is: "No taxation without representation." With technology companies and their algorithms becoming the governing bodies of our modern world, it is again time to fight for the right to be represented, but this time the fight should be focused on representation within the datasets training these systems, the computational models replicating different parts of society, and the workforce building the tools—a fight for "no implementation without representation."

CHAPTER 21

Catalyze Emerging Markets

While corporate operations capable of scaling the globe have been the focus for many years now, what we're beginning to see is that only so many companies can do this. It's impossible for millions of separate companies to dominate the market. Moving forward, it will be more important than ever to figure out ways to accelerate emerging markets and help locally-focused, sustainable businesses participate in the modern economy.

21

The history of Silicon Valley is one filled with stories of outsiders, heretics, and anarchists who started with relatively little and changed the world. Bill Hewlett and Dave Packard started HP out of Packard's garage.[1] Apple was established by two friends tinkering in a garage.[2] Facebook originated in Zuckerberg's dorm room at Harvard University.[3] However, unlike the pioneers who anchored the Valley, today's technologists live extravagant lifestyles. Today, most employees at Big Tech firms have six-figure salaries that come with a signing bonus larger than the annual income of many Americans.[4] They commute from San Francisco to the South Bay on private buses with leather seats, high-speed Wi-Fi, and desks to work on. At work, cleaning and dining staffs make sure everything remains orderly so workers can be as efficient as possible. And then, to celebrate their successes, companies throw extravagant parties filled with models they've paid to be there as part of the ambiance—a scene straight out of Martin Scorsese's film *The Wolf of Wall Street*.[5][6]

But while life is good for many in Silicon Valley, the situation has led to a bubble so big and opaque that many have unknowingly lost touch with the outside world. It has gotten so bad, in fact, that the people working in the Valley have become mentally entrapped by the very circumstances they've created—a state of FOMO in which they believe they have to be in the Valley just to have a shot at doing anything significant. The result, ironically, is that many of the world's most talented technical workers and entrepreneurs are now swimming in a reservoir of stale ideas, chasing after low hanging fruit thanks to the billionaire-fortified walled garden Silicon Valley exists within. As Adam Smith notes in *The Wealth of Nations*, "When you dam up a stream of water, as soon as the dam is full, as much water must run over the damhead as if there was no dam at all."[7]

Create Hyper-Local, Sustainable Businesses

While we idealize the companies established in Silicon Valley, there's much to be learned from companies in smaller markets. For one, distance creates a freedom that is impossible to achieve in a high-pressure environment like that of many companies in the Valley, where employees spend months narrowly focused on a small feature. Getting away from such pressures and allowing our minds to wander often leads to insights that may not have been discovered otherwise. And secondly, success often lies in the fact that someone noticed a specific problem that the rest of the world was incapable of seeing and decided to capitalize on it. Combined, these factors are leading to a rise of innovation outside Silicon Valley, as hyper-local sustainable business models are being created to solve real problems in people's lives, not just get them hooked to a screen.

For example, a company call SALt, from the Philippines, has discovered a way to provide light to people in developing nations by using salt water. And they've figured out how to make it more cost effective than a kerosene lamp.[8] Other companies doing similar things include Avani Eco from the Denpasar area of Bali, Indonesia, which has created an ecological way to replace plastics with industrial-grade cassava starch and other natural ingredients, and Bakeys, a company from Andhra Pradesh in southern India, which has figured out a way to create edible utensils, eliminating the need for plastic.[9][10] Digital examples include Bima, a micro-insurance agency based in Stockholm, focused on supplying insurance to Asia, Africa, and Latin America at rates specific to their market.[11] Others include infarm in Berlin, which is using digital technology to enable urban farming and create local, sustainable food markets, and Think Whole Person Health, which is using digital tech to improve the integrated medical experience of patients in the Omaha, Nebraska metro area.[12][13]

These companies are solving problems that meet local market needs, some of which are easily scalable, others not. Of course, not being able to scale into a billion dollar company shouldn't be viewed as a problem. Markets thrive when companies have competition and are pushing each other to innovate, as Gary Hamel and Michele Zanini of *Harvard Business Review* state in their article titled, "A Few Unicorns Are No Substitute for a Competitive, Innovative Economy."[14] In fact, sometimes competition is needed just to create a marketplace for an idea in the first place, as validated by Elon Musk who has become famous for patenting technology then sharing it publicly.[15] He does this for several reasons. First, he's spent years developing the technology, and his public release of the details is basically a classy way of saying "Here's the secret sauce, good luck trying to catch up." And second, he knows that the best way to get investors interested is not to create a product that exists in a market of its own, but one that draws lots of attention—the more people trying to make it, the more likely investors will think the idea is viable.

By definition, a local, sustainable business isn't meant to make anyone a billionaire. But the world doesn't need more billionaires, it needs more people solving problems in sustainable ways, creating jobs for local economies, and distributing wealth fairly. At a certain point we all have to learn to be happy with making a comfortable living, instead of being focused on being the next billionaire. This isn't settling, it's being realistic, as Eric Ries writes in his book, *The Startup Way*, "We hear a lot about 'unicorn' companies, startups that grow into billion-dollar successes, or even hundreds of billions of dollars. But the truth is, these near-mythical success stories are not what create a continuously evolving system of opportunity."[16] Today it is easier to become a billionaire than a millionaire but the pace we're at is not sustainable, and will inevitably lead to a crash if changes are not made. It's time we get back to being satisfied with living a good life of meaning and sufficient wealth instead of focusing on how to vacuum up every penny on Earth. When business and social incentives align, it's a beautiful thing.

Enable the Long Tail

Disruptive innovation doesn't just hit at the margins, slightly alter an idea, or incrementally redesign a single state of a web view; it radically alters the way we experience life. By remaining focused on finding the peak efficiencies in narrowly-focused issues and low-hanging fruits, Silicon Valley has strategically decided to leave the world-changing opportunities to those on the outside. For these reasons and more, we must recognize that the future of tech will be in discovering and accelerating emerging markets, as a noted by Jack Ma, founder of Alibaba, at the World Economic Forum:

> *The first globalization in human history was controlled by a few kings, and emperors. The last 30 years globalization was controlled by 60,000 big companies. The next 30 years, I bet, we will have [at least] 6 million companies involved in globalization. I'm sure we will make it happen.*[17]

As we discover and accelerate emerging markets around the globe we have to keep in mind that all revolutionary innovations that exist today started small. None were handed to founders on a silver platter. One of the biggest problems modern innovators face is a lack of connection to Big Tech hubs, which results in their efforts going unnoticed and all too often leads to a lack of proper resources. By enabling unique solutions created in smaller markets, we will not only open up opportunity for communities to create new jobs and help others around them but also spur competition and innovation that may not exist otherwise, which is always good for the market.

CHAPTER 22

Reintroduce Stability

At this point in history, we've reached a level of global instability that is unlike anything we've seen in decades. This can be troubling to confront, but what can we do? If we want to avoid any complications, we'll need revolutionary changes to governing systems across the globe. These improvements will need to confront the idealism of Big Tech and create a support system for citizens around the globe to help them through the transition.

22

On December 20, 2017 I had lunch with my good friend and mentor Megan Elliott, director of the new Johnny Carson Center for Emerging Media & Arts at the University of Nebraska. We had both recently gotten back from California—she for studio tours with the new administration, and I for speaking engagements and work meetings. Not long into the lunch we began discussing her experience in San Francisco, about which she said she "saw the future." But unlike most people who visit San Francisco and say they "saw the future," I could tell she had experienced something unsettling by the way she said it. Although she was happy to see various offices and discuss how these companies enable creative working conditions, she also noticed a side of San Francisco that all too often gets ignored: a caste system.

At the top are wealthy Big Tech workers who are capable of taking advantage of the tech welfare system, and live in a world that's all sunshine and trolley cars. In the middle are the natives who now represent a working class being driven into the ground by the machine scheduling of ride sharing services they've been forced to work for if they hope to stay afloat. And at the bottom are the homeless, who have become perpetually destitute, lost to drugs and riddled with mental disease. With this system in place San Francisco, a city known for its ultra-liberal politics, has turned into every liberal hippy's worst nightmare—a private club for the ultra-rich.

Since 2012, housing prices in the Bay Area have been rising at a rate that is more than twice the average of the United States, with some areas growing as fast as 30 percent year-over-year, resulting in a 60 month streak has turned out to be even longer and more dramatic than the streak that came with the dot-com bubble.[1] This has made living in the Bay Area excessively expensive, pushing low-income for a family of four up to $117,400 in 2017, according to the US Department of Housing and Urban Development.[2] These astronomical costs then trickle down into non-tech workers' lives, making it difficult for families to keep up with bills, pulling parents away from their families in order to simply afford basics, and, ultimately, destroying any stability that may remain. Although not as dramatic, similar patterns can be seen in other budding tech communities around the United States, including Los Angeles, California; Boulder, Colorado; Austin, Texas; and more. This is what happens when tech companies move in.

Unfortunately, this imbalance is occurring because of the fact that Big Tech has discovered a way to create a system in which the wealthy are getting wealthier and then using that money to discover creative ways to eliminate the humans who helped them build it in the first place. This is not an anecdotal problem, but one that has been addressed by many global organizations, including the World Bank, which has stated:

> *Many advanced economies face increasingly polarized labor markets and rising inequality—in part because technology augments higher skills while replacing routine jobs, forcing many workers to compete for low-paying jobs. Public sector investments in digital technologies, in the absence of accountable institutions, amplify the voice of elites,*

which can result in policy capture and greater state control. And because the economics of the internet favor natural monopolies, the absence of a competitive business environment can result in more concentrated markets, benefiting incumbent firms.[3]

Accumulation of wealth can be great for a state or nation, but not all paths to wealth are created equal. Sierra Leone is rich in diamonds, but this has become more of a curse than a blessing. In Saudi Arabia oil wealth has been at the root of wars for decades. Now, in the United States, technological innovation is beginning to become a burden to the nation—a vacuum of wealth, that is sucking the life out of local economies and the nation as a whole. What's happening throughout tech communities in the US should be a lesson for the rest of the world to learn from. This is what happens when a nation ignores the well-being of its citizens in favor of wealth accumulation for a select few. If others do not learn from these lessons, they are destined to repeat the process, which would be similarly catastrophic for nations around the globe. What we need now is a policy revolution—not just in the United States, but around the globe—that meets technology companies on their turf and prepares our world for the future that is coming. If these companies are allowed to continue without some form of restraint, communities around the globe will suffer increasing wealth inequality, decreased employment, and general instability.

Create Stable Opportunity and Benefits

Despite Amazon's promises to build jobs and catalyze local economies, it has been proven that the company brings no increase in jobs and actually has a net negative impact on local economies, when combined with the tax breaks and other corporate benefits it receives through bidding wars, which it then abuses to dig the local economy into even deeper debt.[4] For example, despite receiving plenty of corporate cushion, Amazon has over 1,400 employees in Ohio (out of 6,000 total) who rely on the Supplemental Nutrition Assistance Program (SNAP) to buy groceries—a trend happening across every state Amazon's warehouses exists.[5][6] Yet Amazon is so powerful it can get away with this because the company has strategically sucked the life out of the economy at large, which leaves everyone desperate. And it can abuse workers for similar reasons—because machines are ready to take over at a moment's notice. So, want to quit? Go ahead. Then Amazon doesn't have to do the firing, which just makes things easier.

Similar instability is being created by the gig economy, where workers are essentially leased at the desire of companies that own them, a clear example of which is demonstrated by the taxi industry in New York City. Before ride-sharing apps, New York City was capable of regulating the amount of taxi drivers through driver medallions. For decades there were no more than 12,000 to 13,000 drivers because of the cost of acquiring a medallion—up to $1.3 million, less than five years ago.[7] But after ride-sharing apps like Uber and Lyft were introduced to the ecosystem, the number of drivers jumped to 47,000 by 2013 and more than 100,000 less than four years later, not long after which the value of a medallion plummeted to below $200,000.[7] This pushed drivers out of the industry and deep into debt trying to pay off the medallions that once afforded them fair economic opportunity.

Doug Schifter, a former New York City taxi driver, felt the situation had gotten so bad that on February 5, 2018 he decided to commit suicide in front of City Hall to bring attention to the situation he and other drivers have been forced into.[8] In his final post to Facebook, Schifter stated that when he first started, he was able to work a 40-hour week and earn enough to cover living expenses and health benefits. But now, thanks to the gig economy, he was having to work more than 100 hours a week to make a fraction of his original pay and lacked any of the traditional benefits that came with taxi driving, due to employment loopholes ride-sharing companies leverage to avoid paying such fees. Schifter also noted how he had been forced deep into debt trying to pay off his medallion, which now held relatively little value, making it impossible to sell or find other paths to relieve the burden. He signed off stating he would rather die as a sacrifice for awareness than work for such a lousy system and watch as other taxi drivers suffered, too.

Establish a Negative Income Tax Bracket

While the current state of automation and the gig economy is bad, it's only going to get worse if we don't make changes. Luckily, many economists have been attempting to figure out how to restructure the economy to create an economic support system for some time now. The thought, however, from many conservative economists is that while automation will eliminate many jobs, the people who lose their jobs are "unemployed, not unemployable."[9] To remedy this situation, several have suggested creating a reverse income tax system that would require those who keep their jobs to continue paying taxes as normal while those who lose their jobs would be supplied with jobs and health benefits through government initiatives similar to railroad, highway, and other government initiatives of the past. While interesting to consider, there are several fundamental flaws in their theory.

First, there's the fact that capitalism is a theory created before the powerful forms of automation that exist today. While it has worked for several centuries, we now live in a new world where it has been turned inside out. Second, their theory assumes governments, which are already struggling to keep up, can be trusted to determine what jobs should be created—an incredible risk. Finally, this theory assumes that people only work for a paycheck, which is not true for many. If we remove purpose from peoples' lives, bad things are sure to happen, as we saw in farming communities throughout the Midwest throughout the late 1980s and early 1990s when automation cut small farming operations out of the market like a brush fire sweeping over a forest. Those cut out—people who cared for the earth and supplied food for entire communities—were then stripped of a sense of purpose that was difficult, if not impossible, to replicate, leading the number of suicides to

skyrocket. As of 2016 the suicide rate in these communities was discovered to be more than 50 percent higher than it was at its peak during the suicide crisis of the 1980s.[10]

As we eliminate jobs that not only help people put food on their tables and afford healthcare but also give them purpose in life, we will likely see horrendous things happen that could never be logically predicted. And at the speed this change is happening it will surely consume more people than could be predicted as well. For this reason, it's time to consider how we might move beyond capitalism to meet the needs of a future in which human labor will no longer be necessary to alleviate many of our traditional economic burdens.

Offer Universal Resources

In contrast to what conservative economists have been proposing, there exists a more liberal opinion held by many in the tech world—including Jeff Bezos, Elon Musk, and Mark Zuckerberg—who believe we should offer some form of universal basic resources, including money, healthcare, and education.[11] [12] [13] The question we must then ask ourselves is this: *Why would some of the most ruthlessly efficient businessmen in the world believe we need to supply the population with a base level of security?* Perhaps it's because they know what the future holds and fear what might result from a gross disparity in wealth and instability, which, historically, has never favored those at the top. It is more likely, as these gentlemen know, that if people are stripped of the stability in their lives they will revolt. The people living in our modern world may have more domestic luxuries than at any other time in history, but they're still human and, ultimately, will react. Even Citigroup recognized this possibility in its 2005 report titled "Plutonomy: Buying Luxury, Explaining Global Imbalances":

" *Concentration of wealth and spending in the hands of a few, probably has its limits. What might cause the elastic to snap back?*

A threat comes from the potential social backlash...the invisible hands stop working. Perhaps one reason that societies allow plutonomy, is because enough of the electorate believe they have a chance of becoming a Pluto-participant. Why kill off, if you can join it? In a sense this is the embodiment of the 'American dream.' But if voters feel they cannot participate, they are more likely to divide up the wealth pie, rather than aspire to being truly rich.

Could the plutonomies die because the dream is dead, because enough of society does not believe they can participate? The answer is of course yes.[14]

Many people question the idea of universal basic resources, but the model has been proven to work throughout the world. Luxembourg, Singapore, Switzerland, Japan, Australia, Canada and many of the most developed nations around the world have some form of free healthcare, whether complemented by paid options or not.[15] In Germany, Finland, and Norway students, both national and international, pay no tuition to go to school, and it's working.[16] And experiments are being run by Finland, Italy, and Y Combinator in Oakland, CA to understand the potential of universal income.[17][18][19] However, there are also inherent flaws to the idea of universal resources, like the fact that everyone—including the wealthy—would be receiving these benefits, creating even more padding for their cushions

Offer a Mixed Reality

Both sides of this argument have valid points, but what if both sides compromised to create a system that brought the best of both together? As it turns out, this would result in a system similar to that of construction and mining industries, where employees are considered independent contractors and work for several companies throughout the year—a system not so different than the gig economy. With a system like this in place, gig economy employers like Uber, TaskRabbit, and Grubhub, among others, could no longer avoid providing benefits. Such a system may not be what's best for business, but sometimes we have to put a higher value on the people driving the business than the business itself. As Steven Hill, a senior fellow with the New America Foundation, has stated: "Regardless of how many employers a person works for, a worker should not be denied the civilized and modern-day necessity of having access to a support system she needs for herself and her family."[20]

To imagine how such a system might work, we can think of Tom, who provides 20 rides a week for Uber, 10 for Lyft, 30 deliveries for Instacart, and 4 for GrubHub, and Cindy who completed 30 gigs for TaskRabbit and provided 12 rides for Uber between gigs. These two would get their benefits from a centralized source, called individual security accounts (ISAs), which would cover not only basic benefits, but also sick days, paid time off, paid holidays, and other benefits standard in traditional jobs. The money for such funds would be paid by companies on a prorated basis, according to how much work the individual performed for each. This model has been proven to be successful in traditional labor markets and there's no reason to believe it wouldn't work in the modern economy.

With the economy pushing more than one in three workers into independent roles—a number that is expected to surpass one in two within the next five years—and the Supreme Court of the United States deciding to deny workers the ability to file class action lawsuits as of May 21, 2018, figuring out creative ways to reintroduce stability to the market should be considered a necessity.[21][22][23] Similar protections have already been established in other nations around the world, including the European Union, Japan, South Korea, Israel, and Brazil, where they are operating successfully.[24][25] In the United States California and Washington have taken steps to protect workers, but it's time for the rest of the nation to follow suit.[26][27] Once a benefits system is established, we can begin to discuss ways to redistribute wealth.

The key to discussing wealth-distribution programs is to understand that they don't have to be free giveaways. For example, the Bolsa Familia program in Brazil, which has been running for more than 10 years now, gives small cash transfers to families who meet basic requirements that benefit society, including keeping their kids in school and attending regular healthcare examinations. Since implementing this program, Brazil has cut poverty rates in half, reduced the inequality in wealth by 14 percent and raised the likelihood of children being in school by 21 percent.[28] Another route would be to establish a payment for data creation, a job that will more than likely become blue-collar work in the future. A system of portable benefits, incentive-based basic resources, and payment for data creation would fill many of the financial gaps we're seeing in the market today. You may be curious where the money for such resources would come from, given there are so many costs for tax dollars to cover already, but the idea seems much more feasible upon the realization that we still have an untapped pool of wealth to tax: automation.

Tax Automation

Unlike traditional businesses that have no choice but to abide by the laws of the state or nation in which they physically reside, the fluid operations of tech companies can exist anywhere with an internet connection—or, more specifically, where the most beneficial tax laws are. Google, as well as many other tech companies, have been known to operate in European nations to take advantage of relaxed tax laws.[29] In 2017 it was discovered that Apple has been strategically hosting more than $250 billion in cash offshore to avoid taxes.[30] Moreover, Amazon leverages its influence to create bidding wars between local economies that want it to bring jobs to their state.[31] Although bad now, this will only get worse if millions of jobs are lost to automation. If people aren't getting paid, economies will lose billions of dollars in taxable income. And if people don't have money to buy things, economies will lose billions more on taxable land, goods, toll roads, and other things that people would be incapable of paying for. Without action, the trickle-down effects of automation will surely be catastrophic to economies around the world.

To reintroduce stability to economies, taxing automation and the processes that underlie it will become necessary to counterbalance these damages. One of the most well-known people in favor of such a tax is Bill Gates, who, in an interview with Quartz, said, "If a human worker does $50,000 of work in a factory, that income is taxed... If a robot comes in to do the same thing, you'd think we'd tax the robot at a similar level." He continued by explaining that the money generated from this tax would not only create wealth for economies, but also reduce the financial benefits of employing robots, which might slow the pace of automation and, potentially, give us more time to adjust to the new market.[32] While Gates is right that robots should be taxed, they will need to be taxed more than the average individual capable of working

40 hours a week (totaling 2,080 hours per year). Unlike humans, robots are capable of working 24 hours a day, 7 days a week, 365 days a year (totaling 8,760 hours) and should be taxed accordingly.

But tax considerations should not stop there, as robots only represent one form in which automation will come to life. Even more threatening are automated software services, which are magnitudes cheaper to create, maintain, and replicate. But how do we tax systems that are invisible and only operate within the mind of a machine? To collect taxes on these systems, which are really being trained and operated by billions of unpaid laborers, we should consider taxing data collection and processing. A tax such as this would be fluid and unavoidable, meeting tech companies where they exist—in the cloud—and would bring many benefits to the economy, as noted by the Executive Office of the Obama Administration, in their report titled, "Artificial Intelligence, Automation, and the Economy":

> *Tax policy plays a critical role in combating inequality, including income inequality that may be exacerbated by changes in employment from AI-based automation... Progressive taxation is critical for raising adequate revenue to fund national security and domestic priorities, including supporting and retraining workers who may be harmed by increased automation, and will only grow more important if outsized gains continue to accrue at the top while other workers are left struggling.*[33]

As opposed to income, which operates on a fixed, bracket system, a data collection and processing tax could operate on a sliding scale the same way we gauge water and electricity, since data is such a fluid, traceable resource. To ensure collection and processing levels are reported fairly, auditors—such as those appointed to audit algorithms as discussed in Chapter 16—could also be tasked with tracking and measuring what data is running through the system. To ensure small businesses remain competitive and capable of spurring innovation, it is critical that we establish a threshold that allows those under a certain level to freely operate without having to pay this tax, also similar to the threshold suggested in Chapter 16. However, after reaching this defined minimum all operations should be forced to pay a tax proportional to their level of data collection and processing, relative to the amount of data being collected and processed in the market. Creating a taxation system like this would enable economies to rightfully collect taxes regardless of the company's physical location, and would also introduce benefits beyond the money generated.

For instance, it would allow businesses to continue operating as normal while incentivizing both fair competition and sharing of data, because if companies choose to continue monopolizing data by acquiring businesses for their data assets they would have to pay a higher tax. Such a tax would incentivize ethical levels of data collection and processing because businesses focused on mining as much data as possible would have to pay a higher tax. It would enable us to create a safety net that could fund the resources necessary to help people bridge the gap that is coming, forcing those driving the unemployment to pay for the safety net and incentivizing them to slow down their rabid absorption of the market. Finally, if or when we successfully bridge that gap, the revenue generated from this tax could then become a long-term form of social security that replaces the current system, which is quickly disappearing. As we race toward the perfect efficiency of labor, where humans are no longer necessary to the equation, it is becoming increasingly necessary to discover creative ways to reintroduce stability, and I believe a data collection and processing tax is one that should be seriously considered.

Disrupt the Disruption

As echoed in Silicon Valley's rally cry, "move fast and break things," the Valley set out to disrupt the world. Unfortunately, it appears that they have achieved their goal. The gig economy has eaten away at the foundations of fair economic opportunity around the world. Artificially intelligent systems are eliminating millions of jobs at pace and scale we've never experienced before, which is only supposed to get worse over time. And soon, because of this, policymakers will be forced to determine the best path forward or risk leaving their constituents in a dangerous position. Despite this, tech companies are determined to continue forward, ruthlessly forcing their vision on billions of unwilling or unknowing participants. Historically we've referred to people like this as "tyrannical," "authoritarian," or at the very least, "unjust." In tech it's called "disruption"—a word that, by definition, implies instability—and it's just part of business. Unfortunately, if this situation is not managed well it will more than likely result in years, if not decades, of chaos as companies play tug-of-war with the populations involved. This is something even the Standard & Poor (S&P) is concerned about:

> *A degree of inequality is to be expected in any market economy. It can keep the economy functioning effectively, incentivizing investment and expansion — but too much inequality can undermine growth.*

> *Higher levels of income inequality increase political pressures, discouraging trade, investment, and hiring. Keynes first showed that income inequality can lead affluent households (Americans included) to increase savings and decrease consumption, while those with less means increase consumer borrowing to sustain consumption...until those options run out. When these imbalances can no longer be sustained, we see a boom/bust cycle such as the one that culminated in the Great Recession.*

> *Aside from the extreme economic swings, such income imbalances tend to dampen social mobility and produce a less-educated workforce that can't compete in a changing global economy. This diminishes future income prospects and potential long-term growth, becoming entrenched as political repercussions extend the problem...*

Our review of the data, as well as a wealth of research on this matter, leads us to conclude that the current level of income inequality in the U.S. is dampening GDP growth, at a time when the world's biggest economy is struggling to recover from the Great Recession and the government is in need of funds to support an aging population.[34]

However, it doesn't have to be this way. We do have options, and we can look to history for assistance in devising a plan to reintroduce stability into people's lives and empower nations to successfully traverse this revolution. Some of the suggestions addressed in this chapter may sound radical or idealistic, but we have to remember that this new economy was created by idealists who have made radical change. If we hope to react to what they've created, we'll need similar idealism that meets radically new needs. Many of the ideas discussed in this chapter are already taking root and seeing success around the world, and by taking the appropriate steps forward we can implement these plans on an even larger scale to help the world flourish.

CHAPTER 23

Enable Informed Regulation

The way things have been going for Silicon Valley over the past several years, it is only a matter of time before regulation is established. Whether that regulation is created by informed policymakers and citizens or not is a decision they will have to make for themselves.

Although regulation is not a topic anyone wants to talk about, there comes a point at which it becomes necessary. This is true of all industries but especially true of the tech industry, which, despite its size, is still maturing. Consider, for example, that there are more than 2.1 billion people on Facebook; more than 2 billion people using Google's Android operating system in addition to more than one billion using Maps, YouTube, Chrome, Gmail, Search, and Play, individually; and Amazon now serves more than 310 million Prime accounts.[1][2][3][4] Compare these numbers to historical empires and what is discovered is that there is no comparison. The Roman Empire had a population of 50 to 70 million individuals, the Mongol Empire ruled more than 110 million, and the British Empire controlled around 530 million.[5] In terms of religions, the Catholic Church claims nearly 1.3 billion people worldwide, Hinduism just under 1.1 billion, and Judaism just under 17 million.[6][7][8] And unlike these empires which took hundreds or thousands of years to form, Big Tech has reached this size in less than 30 years, giving a select few great power in record time.

In this short time period these companies have reinvented the way the world operates and, in doing so, have built fortunes beyond comprehension. While they deserve to make fortunes for the work they've done to develop the infrastructure necessary to operate as a global society, their power should not remain unchecked. The world now relies on these technologies, and because of this, the population has effectively been corralled into a state of gamified obedience training. Like a dog with a shock collar where limitations are not visible to the eye but learned over time through pain and fear, many citizens have begun to give up any hope of trying to push beyond their invisibly defined limitations. There is no reason this should be happening. This is not leadership, it is an abuse of power. So while it is important that we recognize the incredible value these companies have brought to the world, it's time we make change—it's time we regulate. And if we think about how this might happen, there are really only three ways this plays out, one of which is far superior to the other two.

1. No regulation is put in place, and companies continue functioning as they have been. If regulation remains light and flexible, as it currently is, this means we've decided to leave the decisions to publicly traded corporations that have access to more information than any entity in the history of the world besides perhaps the CIA, the FBI, or other government intelligence agencies. This will undoubtedly lead to an unmonitored arms race similar to the nuclear arms race between the United States and the Soviet Union. But this time it will be focused on creating the most intelligent, most powerful artificially intelligent systems, in pursuit of corporate profit. This is something Elon Musk, a world leader in artificial intelligence who has financial incentives to lead the way, has commented on several times, despite the fact that regulation could harm his companies' financial potential. In an interview at the MIT AeroAstro Centennial Symposium Musk spoke about the need for regulatory oversight, stating:

1 People freak out when Twitter or Netflix go down, but if Google went down for a day or a week (or longer) how much would it impact your life? Would you be capable of working? If you were lost, would you be able to find your way? In general, would you be capable of getting by as you do every day?

Largest Empires in the History of the World

The graph below shows the number of people in the largest empires in the history of the world, compared to the number of years it took for the empire to reach it's peak size.

- ■ Number of people in empire (in billions)
- ▨ Number of years empire existed

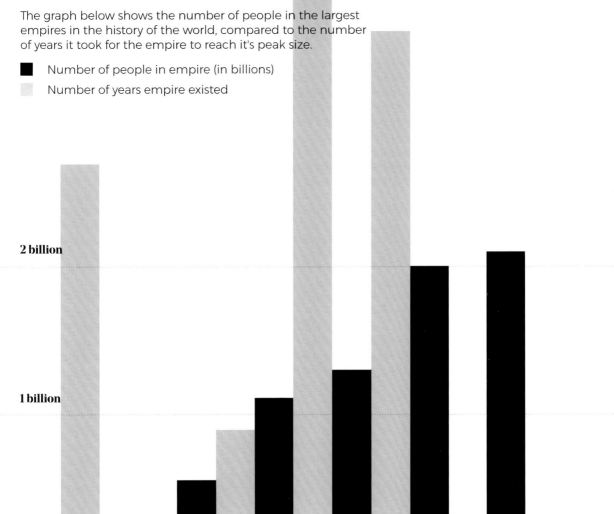

Roman Empire 0.07 B 1,500 Y	**Mongol Empire** 0.11 B 162 Y	**British Empire** 0.53 B 500 Y	**Hinduism** 1.1 B 6,000 Y	**Catholicism** 1.3 B 2000 Y	**Android** 2 B 10 Y	**Facebook** 2.1 B 14 Y

2 billion

1 billion

> *I think we should be very careful about artificial intelligence. If I had to guess at what our biggest existential threat is, it's probably that. So we need to be very careful with artificial intelligence.*
>
> *I'm increasingly inclined to think that there should be some regulatory oversight, maybe at the national and international level, just to make sure that we don't do something very foolish. With artificial intelligence we're summoning the demon.*[9]

What many people reading this book probably don't remember, because we were not alive, is that the nuclear arms race put both the United States and Soviet Union in a position that the military referred to as mutually assured destruction (MAD)—a point at which if either side were to continue onward would result in a disastrous situation for all parties involved.[10] Eventually discussions resulted in safeguards that ensured military restraint, but the same has yet to be seen with the tech industry. Without some form of restraint the powers of modern data empires will undoubtedly come to outweigh the power of any nation-state, country, or previous empire, if they don't already. This, like the nuclear arms race, will become unsafe for everyone involved, both the general public and the people creating these systems.

2. Regulation is determined by those who don't fully understand what's going on. As seen in recent elections, people in countries like the United States, the United Kingdom, France, Germany, Brazil, Spain, and many other developed nations, are uncomfortable with what is happening but remain incapable of creating meaningful change because they lack an understanding of what's going on. This isn't to say that any of these people are dumb, but that they live different lifestyles in which they have no immediate need to understand artificial intelligence. Yet despite this, decisions being made in Silicon Valley and tech communities around the world are disrupting their lives. And in an attempt to restore the order they once knew, these individuals are willing to do whatever it takes to protect their freedoms. Unfortunately, this is bound to lead to regulation that stifles innovation and puts us all in harm's way, from both an economic and military perspective.

From an economic perspective, we have to remember that these systems are fluid and elusive, allowing companies to operate anywhere that has a good Wi-Fi connection. Because of this, overregulation tends to drive innovators to places where regulation is not as strict. And without these innovators countries are destined to fall behind economically. But while falling behind economically would be bad, there should be even greater concern from a military perspective. Falling behind in the artificial intelligence race will quickly leave nations so far behind they'll be unable to secure themselves in times of threat. This would only make the threats from those who wish to use the technology to own the world even greater, as Vladimir Putin's chilling opinion hinted at in 2017:

> *Artificial intelligence is the future, not only for Russia, but for all humankind. It comes with colossal opportunities, but also threats that*

are difficult to predict. Whoever becomes the leader in this sphere will become the ruler of the world.[11]

Recognizing this, we must maintain the belief that trying to shut down these systems out of fear is not a reasonable option. Instead, we have to embrace this future and move forward with courage by working to create a governing strategy that ensures the safety of everyone involved. We need powerful nations to align, set standards for the world, and support each other from the threat of any nation that would rather wield these technologies for power and destruction. Modern regulation should focus on creating fluid, adaptable systems that promote public safety and reinforce corporate responsibility while still allowing companies to flourish economically. Harsh punishments may make constituents happy momentarily, but visceral emotional reactions focused on satisfying immediate needs without fully comprehending the long-term effects are not what we need right now. Instead, we need to focus on the long-term well-being of our world.

3. We accept that regulation is necessary, get involved in our communities, and discuss the options. The difference between previous revolutionary technologies and those being discussed today is that previous revolutionary technologies were often built and controlled by the government, while today's are being built and controlled for commercial purposes.[12] Proof that the technical prowess of Silicon Valley has come to outweigh that of the government, can be seen in the work being done to help advance government defense initiative. While this may concern some, it is undeniable that we need these companies, which possess some of the top talent in the world, to help. This is something companies are aware of, and are beginning to embrace, as noted by Sergey Brin in Alphabet's 2017 Founder's Letter, released in early 2018:

> *Technology companies have historically been wide-eyed and idealistic about the opportunities that their innovations create. And for the overwhelming part, the arc of history shows that these advances, including the Internet and mobile devices, have created opportunities and dramatically improved the quality of life for billions of people. However, there are very legitimate and pertinent issues being raised, across the globe, about the implications and impacts of these advances. This is an important discussion to have. While I am optimistic about the potential to bring technology to bear on the greatest problems in the world, we are on a path that we must tread with deep responsibility, care, and humility.*[13]

Beyond any flaws that may immediately be on your mind after reading what you have thus far, it should also be recognized that at the end of the day there are humans inside these companies. These are people who have families to take care of, lives to build, and a desire to make the world a better place. Most of these people are not evil, and in fact, most of them joined because of the potential to positively impact the world, as discussed in Part 3. That being said, even with the best intentions mistakes can happen. That's part of being human. The difference between the mistakes these people make and the mistakes the average person makes is that those from Silicon Valley are magnified due to the size of their impact—and rightfully so. However, there also are things that make these mistakes difficult to avoid at times—things that aren't easy to understand from the outside.

For example, working with budgets that are larger than the wealth of many nations combined may sound awesome, but is not as glamorous as it sounds. Conversations about making changes are not so different than diplomats discussing national or international budget changes. Toss in some strict deadlines and a dose of global impact, and what you have is bunch of people who are asked to make the impossible possible, and to have it done by tomorrow morning, without any mistakes. Combine this with many other factors that can only be experienced by being on the inside, including not being able to talk to anyone about what you do, and what results is an overwhelmingly stressful work environment. I don't say these things to make excuses or to cast doubt on the capabilities of the talent within these organizations, but to recognize that at the end of the day, these are humans, and they deserve empathy as well.

So, as opposed to how option three sounds, it isn't about giving away intellectual property or helping the general public overthrow anyone. It is about helping the population understand the situation well enough that they can not only empathize with those involved but also feel safe about their future, and the future of their children. It's about figuring out a way we can work together to discover a solution. And, more than anything, it is about regaining the public's trust. There is hard work to be done if we want to balance the matters of national security with personal security and consumer safety, but only by being transparent and helping the public understand what's going on will this happen.

Regardless of whether these corporations want to initiate these conversations or not, every individual on this planet deserves to understand how to remain safe and prosper in the future we are collectively creating. These tools can't be forced on the world without allowing people to understand how they will impact their lives. Helping the general public understand these systems, trust the people behind them, and bridge the gap that is sure to come from an increased use of artificial intelligence should be considered a matter of basic human decency, not a last line of defense to put out PR fires. It's time to bring these issues to light so we can have a larger discussion about what's happening and collectively figure out a way to make it work for everyone, not just a select few.

Create the Future

Now that you're fully educated and aware of the importance of your role in the future of technology, please take the time to read and sign this pledge to show your intent to implement what you've learned.

The Pledge

Change makes people uncomfortable. This is a fact that has been proven time and again in the social sciences, with one of the most commonly cited studies being the one performed by Solomon Asch in 1951. What Asche discovered was that even when the correct response is clear we will conform to the group in order to avoid ridicule.[1] The results of his study demonstrate that we have evolved to take the path of least resistance, and with the infinite complexity of life, it's easy to understand why. People fear change because it means moving forward into the unknown, risking severe instability and potential failure. And at the speed we're moving, it's easy to understand why so many people would prefer the opposite. Trying to keep up with modern technology is like sprinting up a mountain that's falling out from underneath of us as we climb—there's no stopping to catch your breath, and there's no turning back. But as history shows, nothing is known until it happens. There's no way to connect dots that haven't been drawn.

The period we're entering is one that will reveal truths of humanity we've never been capable of realizing before. Approaching these topics will be scary at times. The transparency that has come with this era will force us to look at our truths and our flaws, and ask ourselves questions we may never have considered before. But running away to another planet or attempting to find safe haven in a digital world—some alternate reality—will not solve any of the problems that currently plague us. As long as there are humans involved, there will be no escaping the fragile nature of our humanity.

While there are no general remedies to rid ourselves of the problems we're facing today, the only way we can make any change is by trying. I chose the title "Automating Humanity" not as a scare tactic, but because I believe the title represents the two sides of the world we're currently living in. On one side, there's the fact that we are automating the most intimate aspects of human life into machines. But on the other, there's the fact that while these systems have the potential to become ruthlessly destructive, there's still plenty of time to make sure they're created in humane ways. What you have read in this book is my attempt to outline how we can actionably create a safe, inclusive, and meaningful future, through technology. These suggestions should by no means be considered final destinations to be crystallized in history, but I do hope that by outlining complex issues in a simple way this book has sparked ideas in your mind that you may not have had otherwise. And I hope that those thoughts initiate conversations that might not have started otherwise, because I know that sometimes without a place to start they never begin.

If you enjoyed reading this book, believe in what you've read and plan to help make the changes that need to be seen, please share it with others, and take the time to read and sign this pledge to show your intent. And if you know someone who might enjoy this but can't afford to buy the book themselves, I hope you share your copy. Just don't tell my publisher I said that, they wouldn't be happy.

A Better Future, Together

I, _____, hereby pledge not to use, or stop using, any system that represents or advocates unethical use of technology, and will do what I can to make sure others understand why the tool should be avoided.

I pledge to fight for fair representation within the data sets being used to train systems, the computational models being created to represent segments of society, and the teams I work on, not only because it's the right thing to do but also because it's what's best for business.

And finally, I pledge not to distort, delay, or withhold any information that may skew others' perception of the world, create destructive patterns within automated systems, or otherwise negatively impact the world we are collectively creating.

By taking this pledge I am committing to creating a better future, through technology, which I will help create by leveraging the knowledge I now possess to educate those who may not understand.

SIGNATURE

_____

If you're passionate about these issues, please share your pledge with others. Post it on your social media accounts, send it out in your newsletter, and otherwise bring attention to it how you may. Each pledge is a step towards the future we all deserve to be a part of.

For real-time information, tools, and other resources to help make change, visit: www.designgood.tech

ACKNOWLEDGEMENTS

What It Takes to Make It

Nobody has ever done anything alone. Inspiration, whether it's recognized or not, comes from everywhere. Life experiences change the way the idea comes to life. Conversations alter our perception of the world. Although one individual brain may conceptualize an idea away from the noise of society, nothing was ever created in an isolated bubble. And no matter how much anyone wants to tell you otherwise, this will always be true. These are the people who helped make this happen.

If I've learned anything from this process, it's that it really does take a village to make something special. My name might be on the front cover, but there are so many people who deserve to be recognized. The names listed below do not represent an exhaustive list of everyone that's been involved along the way. I owe so much to so many people for believing in this and helping me make this happen, but the following list does cover a large majority who played a significant role, whether they recognize it or not.

For all of their help getting this book to the finish line, I literally wouldn't have been able to create this book without my good friend and partner in crime, Jessie Sohpaul, who worked tirelessly to make this book cover one of the most beautiful things I've ever seen. I know I was super picky and pushed on things that you might have thought nobody would notice but the face turned out beautiful, man, and the custom cover font matches so well. I'm incredibly proud of the time and effort you gave this work!

My editor Ellen Kleiner, who worked long, tiring hours on an incredibly short schedule just to get this out on time. I know it was frustrating at times, as your lifetime of editorial experience met my first-time writer knowledge, but I appreciate all of the patience, care, and classical training you brought to this project. Your work took this book from well-written and engaging to professional and ready for press. I know how much it took for you to get this done and I'm eternally grateful for what you have done. I can only hope the world appreciates your work as much as I do.

Chelsea Klinginsmith, a good friend of mine who came on at the final hours of production to help us edit extra content that needed to be flipped in an incredibly short amount of time. I handed everything off as we were sprinting to the finish line and you picked it up in stride like you'd been along for the ride the whole time. Thank you so much for your effort, I can promise you we all appreciated it!

And lastly, to powerHouse Books. I spoke with hundreds of people during the process of writing this book, many of whom wanted to introduce me to publishers. I listened to all of their offers, but knew I wasn't going to take anyone's offer until I found the a partner that wanted to help me produce this book with purpose—not for some sensationalized headline and a quick buck—and I'm glad I did. I couldn't be more grateful for your offer to take a risk on me as I am certain I could not have found a better team to assist me in this process.

For all of their help making sure the content was approachable and engaging—whether that meant reading rough drafts, one-on-one discussions to help me think through ideas, or sending articles and other pieces of information to make sure I was up to date with what was happening, this book wouldn't be the same without: Bobby Schulz, Corban Baxter, Rachel Beisel, Hugh Dubberly, Sudha Jamthe, Graham Hill, Shivaun & Adam Raff, Max Miner, Randall Gordon, Adam Jones, Megan Elliott & Brendan Harkin, Rob Van Kranenburg, Chris Messina, Caio Calado, Anamita Guha, Roberto Oliveria, Lauren Mosenthal & Simon Taranto, Brad Jacobson, Aydoğan Schosswald, Dana Holdt & Mario Pesendorfer, Keith Harper, Timóteo Brasil & Keyla Macharet, Rich Gioscia, Kira Tschierschke, John

Labaree, Andreas Santanilla, Pablo Alejo, Deborah Kay, Kate Compton, Bernardo Huberman, Len Beasley, Blake Hudelson, Adam Wagler, Jane Crofts, Davis Godbout, Christian Nitu, Eric Singhartinger, Nick Clement, Maddie Boatwright, Mitchell Clark, Sophie Cummings, and JP Taxman, as well as many others who requested to remain anonymous to avoid potential repercussions due to the controversial nature of the book.

For all of their help connecting me to people around the globe, getting onto podcasts, and generally helping me get the word out there, I would not have gotten the word out for this book without the help of: Rob Van Kranenburg, Leah Taylor, Justin Kraft, Alex Bogusky, Megan Elliott, Sudha Jam the, Dave Burg, Jonathan Schoenberg, Adam Wagler, Paul Glader, Biagio Arobba, Caio Calado, Mahoney Turnbull, Jesse Weaver, Carlos Estrada, Amy Struthers, John M. Fulwider, Brian Ardinger, Timo Hahn, Kristin Hillery & Todd Nienkirk, Fabian Ziegler, Matt O'Donnell, Philip Watson, Elijah Woolery, Andi Galpern, Andrew J. Turner, Rich Gioscia, Simon Morel, Paula Høiby, Cheyenne Noel, Mitchell Clark, and Kelly Lopez.

For all of their help giving me a place to stay, a place to work, or otherwise just generally helping me get through the tough times as I spent more than a year on the road—many times living out of my car, I literally wouldn't have been able to complete this journey without: Rachel & Kalan Beisel, Stefan Knott, Sam Schanfarber, Corban Baxter, Rod Ekwall, Lisa Scala, Benton & Laurel Rochester, Scott Laner, Caio Calado, Luiz Ottino, Jessie Sohpaul, Chris Browne & Jackie Alvine, Taymar & Max Pixleysmith, Timo Hahn, Francisco Martinez, Lauren Mosenthal & Simon Taranto, Matt O'Donnell, Dana Holdt & Mario Pesendorfer, Jacob Smith & Shanell Sanchez-Smith, K-de & Carlos Prado, Christian Hollweg, Christian Nitu, Lauren Mosenthal & Simon Taranto, Kira Tschierschke, Chris Giersch, Brad Ambrose, Stephanie Hall, Julio Soria, Taylor Hendrix, Stephanie Hall, Hailey Buddenberg & Alex Wright, Kat Ringenberg, JP Taxman, Philip & Maryrose Watson, Neal (and Andrea) Thomas, Luke Trevino, Ross Heidenreich, Kevin Bush, TJ Nichols, Hans Christensen, Chris Le, Mary Lemmer, and Antonio & Tracy Aguero.

Learn More

I owe much of this knowledge to authors and writers who have written about this topic long before I did. If you're interested in learning more about their work or discovering where you can read more, in general, this is the place.

End Notes

Chapter 1

1 – John Maynard Keynes, "Economic Possibilities for our Grandchildren," in *Essays in Persuasion* (New York: W. W. Norton & Company, 1963), 358–73, https://www.marxists.org/reference/subject/economics/keynes/1930/our-grandchildren.htm (accessed April 8, 2018).
2 – "Number of Facebook employees from 2004 to 2017 (full-time)," Statista – The Statistics Portal, accessed January 19, 2018, https://www.statista.com/statistics/273563/number-of-facebook-employees/.
3 – "World's Largest Public Companies — Walt Disney," Forbes, accessed January 19, 2018, https://www.forbes.com/companies/walt-disney/.
4 – "Number of Tesla employees from July 2010 to December 2017," Statista – The Statistics Portal, accessed January 19, 2018, https://www.statista.com/statistics/314768/number-of-tesla-employees/.
5 – "Sustainability Report 2017/18," Ford Motor Company, accessed April 9, 2018, https://corporate.ford.com/microsites/sustainability-report-2017-18/people-society/our-people/index.html.
6 – "Number of Amazon employees from 2007 to 2017," Statista – The Statistics Portal, accessed January 19, 2018, https://www.statista.com/statistics/234488/number-of-amazon-employees/.
7 – "Company Facts," Walmart, accessed January 19, 2018, https://corporate.walmart.com/newsroom/company-facts.
8 – Dave Edwards and Helen Edwards, "There are 170,000 fewer retail jobs in 2017—and 75,000 more Amazon robots," Quartz, December 5, 2017, https://qz.com/1107112/there-are-170000-fewer-retail-jobs-in-2017-and-75000-more-amazon-robots/.
9 – Jason Goldberg (@retailgeek), Tweet, March 20, 2017 (1:58 p.m.), https://twitter. com/retailgeek/status/843899612977029121.
10 – Emily Jane Fox, "How Amazon's new jobs really stack up," CNN Money, July 30, 2013, http://money.cnn.com/2013/07/30/news/companies/amazon-warehouse-workers/index.html.
11 – "Smart Diagnosis," LG, accessed Februrary 3, 2018, http://www.lg.com/us/support/answers/front-control-top-load-washers2013/smart-diagnosis.
12 – "Connected Appliances," GE, accessed February 3, 2018, http://www.geappliances.com/ge/connected-appliances/.
13 – RJ Reinhart, "Most Americans Already Using Artificial Intelligence Products," Gallup, March 6, 2018, https://news.gallup.com/poll/228497/americans-already-using-artificial-intelligence-products.aspx.
14 – Gallup and Northeastern University, *Optimism and Anxiety: Views on the Impact of Artificial Intelligence and Higher Education's Response*, January 2018, 9, https://news.gallup.com/reports/226475/gallup-northeastern-university-artificial-intelligence-report-2018.aspx.

Chapter 2

1 – Brief for Data & Society Research Institute and Fifteen Scholars of Technology and Society in Support of Petitioner as Amici Curiae Supporting Respondents, Timothy Ivory Carpenter v. United States of

America, 585 U.S. (2017) (no. 16–402), https://datasociety.net/pubs/fat-ml/DataAndSociety_CarpentervUS_Amicus_Brief.pdf.

2– "Mobile Fact Sheet," Pew Research Center, accessed January 28, 2018, http://www.pewinternet.org/fact-sheet/mobile/.

3 – "Privacy," Apple, accessed January 18, 2018, https://www.apple.com/privacy/.

4 – "Multitasking: Switching Costs," American Psychological Association, accessed January 22, 2018, http://www.apa.org/research/action/multitask.aspx; Ira Flatow, "The Myth of Multitasking," NPR, May 10, 2013. http://www.npr.org/2013/05/10/182861382/the-myth-of-multi-tasking; Adam Gorlick, "Media Multitaskers Pay Mental Price, Stanford Study Shows," Stanford, August 24, 2009, http://news.stanford.edu/news/2009/august24/multitask-research-study-082409; Gigaom, "Clifford Nass: Multi-Tasking is Bad for Your Brain," YouTube video, 7:54, Uploaded on May 28, 2013, https://www.youtube.com/watch?v=BEb-mUQpwR2E.

5 – Robert Tercek, *Vaporized: Solid Strategies for Success in a Dema-terialized World*, (Vancouver, BC: LifeTree Media, 2015), 63.

6 – Josh Constine, "Facebook changes mission statement to 'bring the world closer together'," *Techcrunch*, June 23, 2017, https://techcrunch.com/2017/06/22/bring-the-world-closer-together/.

7 – Eric Rosenberg, "The Business of Google (GOOG)," Investopedia, accessed February 18, 2018, https://www.investopedia.com/articles/in-vesting/020515/business-google.asp.

8 – "Facebook's advertising revenue worldwide from 2009 to 2017 (in million U.S. dollars)," Statista – The Statistics Portal, accessed February 8, 2018, https://www.statista.com/statistics/271258/facebooks-advertis-ing-revenue-worldwide/.

9 – Sarah Perez, "Amazon launches a 'lite' Android web browser in India," Techcrunch, April 18, 2018, https://techcrunch.com/2018/04/17/amazon-launches-a-lite-android-web-browser-app-in-india/.

10 – Daniel Van Boom, "Why India snubbed Facebook's free Internet offer," cnet, February 26, 2016, https://www.cnet.com/news/why-india-doesnt-want-free-basics/.

11 – Shelly Walia, "Facebook board member Marc Andreessen: Indians should've embraced Free Basics—and colonialism," Quartz, February 10, 2016, https://qz.com/613815/facebook-board-member-marc-an-dreessen-indians-shouldve-embraced-free-basics-and-colonialism/

12 – Mark Zuckerberg, "The technology behind Aquila," Facebook, July 21, 2016, https://www.facebook.com/notes/mark-zuckerberg/the-tech-nology-behind-aquila/10153916136506634/.

13 – Tercek, 131.

Chapter 3

1 – Nir Eyal, *Hooked: How to Build Habit Forming Products* (New York: Portfolio/Penguin, 2014), 13.

2 – Eyal, 34.

3 – "Addictions," American Psychological Association, accessed Febru-ary 22, 2018, http://www.apa.org/topics/addiction/.

4 – Justin Lafferty, "STUDY: How Addicted Are We To Facebook Mobile?," *Daily News*, March 27, 2013, https://www.adweek.com/digital/facebook-idc-study-smartphones/.

5 – Golden Krishna, *The Best Interface Is No Interface: The Simple Path to Brilliant Technology* (San Francisco: New Riders, 2015), 92.

6 – Meena Hart Duerson, "We're addicted to our phones: 84% worldwide say they couldn't go a single day without their mobile device in their hand," *Daily News*, August 16, 2012, http://www.nydailynews.com/life-style/addicted-phones-84-worldwide-couldn-single-day-mobile-device-hand-article-1.1137811.

7 – Zoran Milich, "One-third of Americans would give up sex for their mobile phones - survey," *RT News*, January 16, 2015, https://www.rt.com/usa/223335-sex-mobiles-survey-americans/.

8 – Mike Allen, "Sean Parker unloads on Facebook: 'God only knows what it's doing to our children's brains,'" *Axios*, November 9, 2017, https://www.axios.com/sean-parker-unloads-on-facebook-god-only-knows-what-its-doing-to-our-childrens-brains-1513306792-f855e7b4-4e99-4d60-8d51-2775559c2671.html.

9 – Stanford Graduate School of Business, "Chamath Palihapitiya, Founder and CEO Social Capital, on Money as an Instrument of Change," video, 56:15, November 13, 2017, https://www.youtube.com/watch?time_continue=1286&v=PMotykw0SIk.

10 – Katherine Schwab, "Nest Founder: 'I Wake Up In Cold Sweats Thinking, What Did We Bring To The World?,'" *FastCo Design*, July 7, 2017, https://www.fastcodesign.com/90132364/nest-founder-i-wake-up-in-cold-sweats-thinking-what-did-we-bring-to-the-world.

11 – Benjamin Baddeley, Sangeetha Sornalingam, and Max Cooper, "Sitting is the new smoking: where do we stand?," *British Journal of General Practice* 66 (646): 258 (2016), doi:10.3399/bjgp16X685009.

12 – Amy Cuddy, "Your iPhone Is Ruining Your Posture—and Your Mood," *New York Times*, December 12, 2015, https://www.nytimes.com/2015/12/13/opinion/sunday/your-iphone-is-ruining-your-posture-and-your-mood.html.

13 – "Preventing and dealing with that Hump on the back of your neck," The Posture Clinic, accessed February 21, 2017, http://www.postureclinic.ca/preventing-and-dealing-with-that-hump-on-the-back-of-your-neck/.

14 – Nair Shwetha et al., "Do slumped and upright postures affect stress responses? A randomized trial," *Health Psychology*, 34 (6): 632–41 (2015), doi:10.1037/hea0000146.

15 – David Dudley, "Quest for a Good Night's Sleep," *AARP*, August/September 2016, https://www.aarp.org/health/conditions-treatments/info-2016/sleep-apnea-insomnia.html.

16 – M. A. Christensen et al. "Direct Measurements of Smartphone Screen-Time: Relationships with Demographics and Sleep," PLoS ONE 11 (11): e0165331, (2006), doi:10.1371/journal.pone.0165331; Ji Hye Oh et al., "Analysis of circadian properties and healthy levels of blue light from smartphones at night?," *Scientific Reports* 5: 11325 (2015), accessed March 29, 2018, doi:10.1038/srep11325.

17 – Nikhil Swaminathan, "Can a Lack of Sleep Cause Psychiatric Dis-

orders?," *Scientific American*, October 23, 2017, https://www.scientificamerican.com/article/can-a-lack-of-sleep-cause/.

18 – Rocci Luppicini, *Handbook of Research on Technoself: Identity in a Technological Society* (Hershey, Pennsylvania: IGI Global, 2012), 26-39.

19 – Jean M. Twenge, "Have Smartphones Destroyed a Generation?," *The Atlantic*, September, 2017, https://www.theatlantic.com/magazine/archive/2017/09/has-the-smartphone-destroyed-a-generation/534198/.

20 – Jean M. Twenge, *iGen: Why Today's Super-Connected Kids Are Growing Up Less Rebellious, More Tolerant, Less Happy—and Completely Unprepared for Adulthood—and What That Means for the Rest of Us* (New York: Ataria Books, 2017), 97.

21 – Twenge, *iGen*, 103.

22 – H. B. Shakya and N. A. Christakis, "Analysis of circadian properties and healthy levels of blue light from smartphones at night," *American Journal of Epidemiology* 185 (3): 203–11 (2017), 2018, doi: 10.1177/2167723376.

23 – Jean Twenge, et al., "Increases in Depressive Symptoms, Suicide-Related Outcomes, and Suicide Rates Among U.S. Adolescents After 2010 and Links to Increased New Media Screen Time," *Clinical Psychology Science* 6 (1): 3–17 (2017), doi:10.1177/2167702617723376.

24 – Twenge, *iGen*, 101.

Chapter 4

1 – "The Story of Propaganda," American Historical Association, accessed May 8, 2018, https://www.historians.org/about-aha-and-membership/aha-history-and-archives/gi-roundtable-series/pamphlets/em-2-what-is-propaganda-(1944)/the-story-of-propaganda.

2 – Tim Wu, *The Master Switch: The Rise and Fall of Information Empires* (New York: Vintage, 2011), 210.

3 – "History of the BBC," BBC, accessed May 23, 2018, http://www.bbc.co.uk/timelines/zxqc4wx.

4 – "About ITV," ITV, accessed May 23, 2018, http://www.itvplc.com/about/history/2017.

5 – John McCarthy, "The Home Information Terminal — A 1970 View," Stanford University, June 1, 2000, http://www-formal.stanford.edu/jmc/hoter2.pdf.

6 – Larry Page and Sergey Brin, "The Anatomy of a Large-Scale Hypertextual Search Engine," Stanford, accessed March 23, 2018, http://infolab.stanford.edu/~backrub/google.html.

7 – McCarthy, "The Home Information Terminal — A 1970 View," Stanford University, 7.

8 – "Newspapers Fact Sheet," Pew Research Center, accessed April 2, 2018, http://www.journalism.org/fact-sheet/newspapers/.

9 – William Brady et al., "Emotion shapes the diffusion of moralized content in social networks," *Proceedings of the National Academy of Sciences* 114 (28): 7313–18 (2017), doi:10.1073/pnas.1618923114.

10 – Salesforce Research, *Digital Advertising 2020: Insights into a new*

era of advertising and media buying, February 2018, 8, https://www.salesforce.com/content/dam/web/en_us/www/assets/pdf/datasheets/digital-advertising-2020.pdf.

11 – Cathy O'Neil, *Weapons of Math Destruction: How Big Data Increases Inequality and Threatens Democracy*, (New York: Penguin Books, 2017), 187–88.

12 – "Manipulating Social Media to Undermine Democracy," *Freedom House*, November 2017, https://freedomhouse.org/report/freedom-net/freedom-net-2017.

13 – "Twitter's advertising revenue worldwide from 2014 to 2018 (in billion U.S. dollars)," Statista — The Statistics Portal, accessed April 13, 2018, https://www.statista.com/statistics/271337/twitters-advertising-revenue-worldwide/.

14 –"Facebook's advertising revenue worldwide from 2009 to 2017 (in million U.S. dollars)," Statista — The Statistics Portal, accessed April 13, 2018, https://www.statista.com/statistics/271258/facebooks-advertising-revenue-worldwide/.

15 – "Google's ad revenue from 2001 to 2017 (in billion U.S. dollars)" Statista — The Statistics Portal, accessed April 13, 2018, https://www.statista.com/statistics/266249/advertising-revenue-of-google/.

16 – Alyssa Newcomb, "House Votes in Favor of Letting ISPs Sell Your Browsing History," NBC News, March 29, 2017, https://www.nbcnews.com/tech/security/house-set-vote-whether-isps-can-sell-your-data-without-n739166.

17 – David Carr, "How Obama Tapped Into Social Networks' Power," *New York Times*, November 9, 2008, https://www.nytimes.com/2008/11/10/business/media/10carr.html

18– Sandra Rendgen, *Understanding The World: The Atlas of Infographics* (Cologne, Germany: Taschen, 2014), 196.

19 – Scott Galloway, *The Four: The Hidden DNA of Amazon, Apple, Facebook and Google* (New York: Portfolio, 2017), 164.

Chapter 5

1 – George Orwell, *1984* (New York: Signet Classics, 1961), 19.

2 – Josh Chin, "Chinese Police Add Facial-Recognition Glasses to Surveillance Arsenal," *Wall Street Journal*, February 7, 2018, https://www.wsj.com/articles/chinese-police-go-robocop-with-facial-recognition-glasses-1518004353.

3 – Joyce Liu and Wang Xiqing, "In Your Face: China's All Seeing State," BBC, December 10, 2017, http://www.bbc.com/news/av/world-asia-china-42248056/in-your-face-china-s-all-seeing-state.

4 – Rachel Botsman, "Big data meets Big Brother as China moves to rate its citizens," *Wired*, October 21, 2018, http://www.wired.co.uk/article/chinese-government-social-credit-score-privacy-invasion.

5 – Liu and Xiqing, "In Your Face: China's All Seeing State."

6 – Larry Dignan, "Apple's App Store 2016 revenue tops $28 billion mark, developers net $20 billion," ZDNet, January 5, 2017, http://www.zdnet.com/article/apples-app-store-2016-revenue-tops-28-billion-mark-developers-net-20-billion/.

7 – "App Store Review Guidelines," Apple, accessed May 27, 2018, https://developer.apple.com/app-store/review/guidelines/.

8 – Boris Kachka, "James Patterson: 'If Amazon Is the New American Way, Then Maybe It Has to Be Changed'," *Vulture*, May 29, 2014, http://www.vulture.com/2014/05/james-patterson-calls-out-amazon-at-book-expo.html.

9 – Carolyn Kellogg, "Amazon launches attack on Hachette over e-book pricing'," *Vulture*, accessed August 9, 2014, http://www.latimes.com/books/jacketcopy/la-et-jc-amazon-launches-attack-on-hachette-ebooks-20140809-story.html.

10 – Nick Wingfield and Michael J. de la Merced, "Amazon to Buy Whole Foods for $13.4 Billion," *New York Times*, June 16, 2017, https://www.nytimes.com/2017/06/16/business/dealbook/amazon-whole-foods.html.

11 – Eugene Kim, "Amazon Alexa users buy more stuff, study shows," CNBC, December 6, 2017, https://www.cnbc.com/2017/12/06/amazon-alexa-customers-buy-more.html.

12 – Jerome Ribot, "Cashless Effect," Coglode, accessed April 7, 2018, http://coglode.com/gem/cashless-effect.'

13 – Julie Creswell, "How Amazon Steers Shoppers to Its Own Products," *New York Times*, June 23, 2018, https://www.nytimes.com/2018/06/23/business/amazon-the-brand-buster.html.

14 – Dominic Gates and Angel Gonzalez, "Amazon lines up fleet of Boeing jets to build its own air-cargo network," *Seattle Times*, March 9, 2016, https://www.seattletimes.com/business/boeing-aerospace/amazon-to-lease-20-boeing-767s-for-its-own-air-cargo-network/.

15 – Laura Stevens, "Amazon to Launch Delivery Service That Would Vie With FedEx, UPS," *Wall Street Journal*, February 9, 2018, https://www.wsj.com/articles/amazon-to-launch-delivery-service-that-would-vie-with-fedex-ups-1518175920.

16 – James Curlander et al., Lane assignments for autonomous vehicles, US Patent 9,547,986, filed November 19, 2011, and issued January 17, 2017, https://patents.google.com/patent/US9547986B1/en.

17 – Abha Bhattarai, "Amazon is buying smart-doorbell maker Ring," *Wall Street Journal*, February 27, 2018, https://www.washingtonpost.com/news/business/wp/2018/02/27/amazon-is-buying-smart-doorbell-maker-ring/?utm_term=.819da693421f.

18 – "2018 Letter to Shareholders," Amazon, accessed May 8, 2018. https://www.sec.gov/Archives/edgar/data/1018724/000119312518121161/d456916dex991.htm.

19 – Michelle Castillo, "Amazon is straying from Jeff Bezos' vow to 'charge less'," CNBC, April 26, 2018, https://www.cnbc.com/2018/04/26/amazon-prime-increase-strays-from-bezos-charge-less-promise.html.

20 – Galloway, 165.

21 – Wu Youyou, Michal Kosinski, and David Stillwell, "Computer-based personality judgments are more accurate than those made by humans," *Proceedings of the National Academy of Sciences of the United States* 112 (4): 1036–40 (2015), doi:10.1073/pnas.1418680112.

22 – Adam D. I. Kramer, Jamie E. Guillory, and Jeffrey T. Hancock, "Experimental evidence of massive-scale emotional contagion through

social networks," *Proceedings of the National Academy of Sciences of the United States* 111 (24): 8788–90 (2014), doi:10.1073/pnas.1320040111.

23 – Sam Levin, "Facebook told advertisers it can identify teens feeling 'insecure' and 'worthless,'" *Guardian*, May 1, 2017, https://www.theguardian.com/technology/2017/may/01/facebook-advertising-data-insecure-teens.

24 – Dave Lee, "Facebook: Now for young children too," BBC, December 5, 2017, http://www.bbc.com/news/technology-42232475.

25 – Bryan Clark, "Facebook won the war for your mind. Now it wants your children.," The Next Web, December 7, 2017, https://thenextweb.com/facebook/2017/12/06/facebook-already-won-the-war-for-your-mind-now-it-wants-your-children/.

26 – "Joe Camel Advertising Campaign Violates Federal Law, FTC Says," FTC, May 28, 1997, https://www.ftc.gov/news-events/press-releases/1997/05/joe-camel-advertising-campaign-violates-federal-law-ftc-says.

27 – Danny Sullivan, "Google now handles at least 2 trillion searches per year," Search Engine Land, May 24, 2016, https://searchengineland.com/google-now-handles-2-999-trillion-searches-per-year-250247

28 – "YouTube Company Statistics," Statistic Brain Research Institute, accessed May 18, 2018, https://www.statisticbrain.com/youtube-statistics/.

29 – "Gartner Says Worldwide Sales of Smartphones Grew 9 Percent in First Quarter of 2017," Gartner, March 23, 2017, https://www.gartner.com/newsroom/id/3725117.

30 – "Our Products," Google, accessed May 18, 2018, https://www.google.com/about/products/.

31 – Derek Thompson, "Google's CEO: 'The Laws Are Written by Lobbyists,'" *The Atlantic*, October 1, 2010, https://www.theatlantic.com/technology/archive/2010/10/googles-ceo-the-laws-are-written-by-lobbyists/63908/.

32 – Page and Brin, "The Anatomy of a Large-Scale Hypertextual Search Engine."

33 – Cemal Karakas, "Google antitrust proceedings: Digital business and competition," (briefing, European Parliamentary Research Service, July 2015), http://www.europarl.europa.eu/RegData/etudes/BRIE/2015/565870/EPRS_BRI(2015)565870_EN.pdf.

34 – Josie Cox, "Google appeals record EU fine over 'unfair' shopping searches," *Independent*, September 11, 2017, http://www.independent.co.uk/news/business/news/google-eu-fine-shopping-search-results-unfair-competition-european-commission-a7941036.html.

35 – "Statement of the Federal Trade Commission Regarding Google's Search Practices In the Matter of Google Inc. FTC File Number 111-0163," Federal Trade Commission, January 3, 2013, https://www.ftc.gov/system/files/documents/public_statements/295971/130103googlesearchstmtofcomm.pdf.

36 – Adam Raff and Shivaun Raff, "The European Commission's Google Search Case: A Timeline of Significant Events," Foundem, January 21, 2018, http://www.foundem.co.uk/Foundem_Google_Timeline.pdf.

37 – Page and Brin, "The Anatomy of a Large-Scale Hypertextual

Search Engine."
38 – Natasha Lomas, "Google parent Alphabet forms holding company, XXVI, to complete 2015 corporate reorganization," TechCrunch, September 4, 2017, https://techcrunch.com/2017/09/04/google-parent-alphabet-forms-holding-company-xxvi-to-complete-2015-corporate-reorganization/;
39 – "2016 12 19 PAGA Complaint Against Google," Zerohedge, December 9, 2016, https://www.scribd.com/document/334742749/2016-12-19-PAGA-Complaint-Against-Google.
40 – "Annual Reports," OpenSecrets.org, accessed February 19, 2018, http://www.opensecrets.org/about/reports.php.
41 – Jack Nicas, "Eric Schmidt to Step Down as Executive Chairman of Google Parent," *Wall Street Journal*, December 22, 2017, https://www.wsj.com/articles/eric-schmidt-stepping-down-as-executivechairman-of-google-parent-alphabet-1513894599.
42 – "Eric Schmidt," US Department of Defense Innovation Board, accessed May 12, 2018, http://innovation.defense.gov/Media/Biographies/Bio-Display/Article/1377390/dr-eric-schmidt/.
43 – theyTOLDyou, "Google CEO Eric Schmidt on privacy," video, 00:33, December 8, 2009, https://www.youtube.com/ watch?v=A6e7wf-DHzew.
44 – Glenn Greenwald, "Why Privacy Matters," filmed October 2014 at TEDGlobal 2014, video, 20:38, https://www.ted.com/talks/glenn_greenwald_why_privacy_matters.

Chapter 6

1 – Friedrich Engels, Condition of the Working Class in England, (Panther, 1969), https://www.marxists.org/archive/marx/works/download/pdf/condition-working-class-england.pdf.
2 – "Lowell Mill Women Create the First Union of Working Women," The American Federation of Labor and Congress of Industrial Organizations, accessed May 18, 2018, https://aflcio.org/about/history/labor-history-events/lowell-mill-women-form-union.
3 – "Triangle Shirtwaist Factory Fire," History, 2009, https://www.history.com/topics/triangle-shirtwaist-fire.
4 – Maggie Astor, "Microchip Implants for Employees? One Company Says Yes," *New York Times*, July 25, 2017, https://www.nytimes.com/2017/07/25/technology/microchips-wisconsin-company-employees.html.
5 – Anonymous, "What are the negatives in working in a too-good-to-be-true office like Google?" Quora, April 7, 2016, https://www.quora.com/What-are-the-negatives-in-working-in-a-too-good-to-be-true-office-like-Google.
6 – Charlie Kindel, "Friday was my last day at Amazon," cek.log, April 30, 2018, https://ceklog.kindel.com/2018/04/30/friday-was-my-last-day-at-amazon/.
7 – Heather Boushey and Sarah Jane Glynn, "There Are Significant Business Costs to Replacing Employees," Center For American Progress, November 16, 2012, https://www.americanprogress.org/

issues/economy/reports/2012/11/16/44464/there-are-significant-business-costs-to-replacing-employees/.

8 – O'Neil, 105–7.

9 – Julia Angwin, Noam Scheiber, and Ariana Tobin, "Dozens of Companies Are Using Facebook to Exclude Older Workers From Job Ads," ProPublica, April 30, 2108, https://www.propublica.org/article/facebook-ads-age-discrimination-targeting.

10 – Jodi Kantor, "Working Anything but 9 to 5," *New York Times*, accessed May 11, 2018, https://www.nytimes.com/interactive/2014/08/13/us/starbucks-workers-scheduling-hours.html.

11 – Clare Duffy, "New law requires employers to give workers advance notice of their schedules," *Portland Business Journal*, June 29, 2017, https://www.bizjournals.com/portland/news/2017/06/29/new-law-requires-employers-to-give-workers-advance.html.

12 – Alex Campolo et al, "AI Now 2017 Report," AI Now Institute, October 2017, https://ainowinstitute.org/AI_Now_2017_Report.pdf.

13 – Farhad Manjoo, "The Happiness Machine," Slate, January 21, 2013, 2018, http://www.slate.com/articles/technology/technology/2013/01/google_people_operations_the_secrets_of_the_world_s_most_scientific_human.single.html.

14 – "Dynamic 365," Microsoft, accessed June 3, 2018, https://dynamics.microsoft.com/en-us/talent/overview/.

15 – "Intellego for PeopleSoft Enterprise Human Resources," IBM, accessed June 3, 2018, https://www-01.ibm.com/software/info/channel-solution-profiles/Sirius_PeopleSoft_HR.html.

Chapter 7

1 – "World Robotics 2017" International Federation of Robotics, accessed April 23, 2018, https://ifr.org/downloads/press/Executive_Summary_WR_2017_Industrial_Robots.pdf.

2 – "Industrial Robots - Statistics & Facts," Statista, accessed May 3, 2018, https://www.statista.com/topics/1476/industrial-robots/.

3 – Robert Tercek, 210-211.

4 – Infographic from: Andrew McAfee, "The Rebound that Stayed Flat," accessed June 3, 2018, http://andrewmcafee.org/2012/01/mcafee-recession-recovery-economy-employment-graph/.

5 – PBS NewsHour. "Watch President Barack Obama's full farewell speech" YouTube video, 17:54-18:05. Posted. January 10, 2017. https://www.youtube.com/watch?v=udrKnXueTW0.

6 – Seth Weintraub, "Google acquires reCAPTCHA in two-for-one deal," Computerworld, accessed May 5, 2018, https://www.computerworld.com/article/2467668/e-commerce/google-acquires-recaptcha-in-two-for-one-deal.html.

7 – "Quick, Draw," Google, accessed May 5, 2018, https://quickdraw.withgoogle.com/

8 – John Frank Weaver, "What Exactly Does It Mean to Give a Robot Citizenship?," Slate, accessed May 3, 2018, http://www.slate.com/articles/technology/future_tense/2017/11/what_rights_does_a_robot_get_with_citizenship.html.

Chapter 8

1 – Tim Wu, *The Master Switch: The Rise and Fall of Information Empires* (New York: Vintage, 2011), 36.
2 – Yuval Noah Harari, *Sapiens: A Brief History of Humankind* (New York: HarperCollins, 2015), 24
3 – Harari, 124.
4 – Hellmut E. Lehmann-Haupt, "Johannes Gutenberg — German Printer," Encyclopedia Brittanica, accessed May 3, 2018, https://www.britannica.com/biography/Johannes-Gutenberg.
5 – "Guglielmo Marconi," History, accessed May 3, 2018, http://www.history.com/topics/inventions/guglielmo-marconi.
6 – Wu, 33–4.
7 – Kat Eschner, "The Farmboy Who Invented Television," *Smithsonian Magazine*, August 28, 2017, https://www.smithsonianmag.com/smart-news/farmboy-who-invented-television-while-plowing-180964607/.
8 – Kayla Webley, "How the Nixon-Kennedy Debate Changed the World," *Time*, September 23, 2010, http://content.time.com/time/nation/article/0,8599,2021078,00.html.
9 – "Total number of Websites," Internet Live Stats, accessed May 8, 2018, http://www.internetlivestats.com/total-number-of-websites/.
10 – "Number of internet users worldwide from 2005 to 2017 (in millions)," Statistac – The Statistics Portal, accessed May 8, 2018, https://www.statista.com/statistics/273018/number-of-internet-users-worldwide/.
11 – David Christian, *Maps of Time: An Introduction to Big History* (Berkeley: University of California Press, 2004), 344–5; Angus Maddison, *The World Economy, Vol. 2* (Paris: Development Centre of the Organization of Economic Co-operation and Development, 2001), 636.
12 – Angus Maddison, *The World Economy, Vol. 1* (Paris: Development Centre of the Organization of Economic Co-operation and Development, 2006), 261.
13 – "Gross Domestic Product 2017," World Bank, accessed April 22, 2018, https://databank.worldbank.org/data/download/GDP.pdf).
14 – Harari, 58.

Chapter 10

1 – "Timeline of Computer History," Computer History Museum, accessed April 8, 2018, http://www.computerhistory.org/timeline/1952/.
2 – Joseph Weizenbaum, "Computational Linguistics," Stanford University, accessed April 10, 2018, https://web.stanford.edu/class/linguist238/p36-weizenabaum.pdf.
3 – Robinson Meyer, "Even Early Focus Groups Hated Clippy," *The Atlantic*, June 23, 2015, https://www.theatlantic.com/technology/archive/2015/06/clippy-the-microsoft-office-assistant-is-the-patriarchys-fault/396653/.
4 – Bruce Weber, "Swift and Slashing, Computer Topples Kasparov," *New York Times*, May 12 1997, https://www.nytimes.com/1997/05/12/nyregion/swift-and-slashing-computer-topples-kasparov.html.

5 – John Markoff, "Computer Wins on 'Jeopardy!': Trivial, It's Not," New York Times, February 16, 2011, https://www.nytimes.com/2011/02/17/science/17jeopardy-watson.html.

6 – David Silver et al., "Mastering the game of Go with deep neural networks and tree search," *New York Times*, January 27, 2016, https://www.nature.com/articles/nature16961.

7 – Alan Levinovitz, "The Mystery of Go, the Ancient Game that Computers Still Can't Win," *Wired*, May 12, 2014, https://www.wired.com/2014/05/the-world-of-computer-go/.

8 – Erik Brynjolfsson and Andrew McAfee, *The Second Machine Age: Work, Progress, and Prosperity in a Time of Brilliant Technologies* (New York: W. W. Norton & Company, 2014), 49–50.

9 – Brynjolfsson and McAfee, 49–50.

10 – Gordon E. Moore, "Cramming More Components onto Integrated Circuits," *Proceedings of the IEEE* 86, no. 1 (January 1988): 82–85, doi: 10.1109/JPROC.1998.658762.

11 – "IBM Research Alliance Builds New Transistor for 5nm Technology," IBM, June 5, 2017, https://www-03.ibm.com/press/us/en/pressrelease/52531.wss.

12 – John Markoff, "The iPad in Your Hand: As Fast as a Supercomputer of Yore," *New York Times*, May 9, 2011, https://bits.blogs.nytimes.com/2011/05/09/the-ipad-in-your-hand-as-fast-as-a-supercomputer-of-yore/.

13 – Sergey Brin, "2017 Founder's Letter," Alphabet, accessed May 7, 2018, https://abc.xyz/investor/founders-letters/2017/index.html.

14 – "Long-term price trends for computers, TVs, and related items," Bureau of Labor Statistics, October 13, 2015, https://www.bls.gov/opub/ted/2015/long-term-price-trends-for-computers-tvs-and-related-items.html.

15 – John Gantz and David Reinsel, "The Digital Universe in 2020: Big Data, Bigger Digital Shadows, and Biggest Growth in the Far East," EMC Corporation, February 2013, https://www.emc.com/collateral/analyst-reports/idc-the-digital-universe-in-2020.pdf.

16 – "Twitter Usage Statistics," Internet Live Stats, accessed May 7, 2018, http://www.internetlivestats.com/twitter-statistics/.

17 – "Hours of video uploaded to YouTube every minute as of July 2015," Statista — The Statistics Portal, accessed May 7, 2018, https://www.statista.com/statistics/259477/hours-of-video-uploaded-to-youtube-every-minute/.

18 – "The Top 20 Valuable Facebook Statistics – Updated June 2018," Zephoria, accessed June 7, 2018, https://zephoria.com/top-15-valuable-facebook-statistics/.

19 – M. G. Seigler, "Eric Schmidt: Every 2 Days We Create As Much Information As We Did Up To 2003," TechCrunch, August 4, 2010, https://techcrunch.com/2010/08/04/schmidt-data/.

20 – Gantz and Reinsel, "The Digital Universe in 2020."

21 – Tercek, 130.

22 – Ray Kurzweil, "The Law of Accelerating Returns," Kurzweil AI, March 7, 2001, http://www.kurzweilai.net/the-law-of-accelerating-returns.

23 – Kurzweil, "The Law of Accelerating Returns."

24 – "Internet of Things (IoT) connected devices installed base world-wide from 2015 to 2025 (in billions)," Statista — The Statistics Portal, accessed May 12, 2018, https://www.statista.com/statistics/471264/iot-number-of-connected-devices-worldwide/.

25 – IDC, "The Digital Universe of Opportunities: Rich Data and the Increasing Value of the Internet of Things," EMC, March 2014, https://www.emc.com/collateral/analyst-reports/idc-digital-universe-2014.pdf.

26 – Joseph Bradley, Joel Barbier, and Doug Handler, "Embracing the Internet of Everything to Capture Your Share of $14.4 Trillion," (white paper, Cisco Systems, February 12, 2013), https://www.cisco.com/c/dam/en_us/about/ac79/docs/innov/IoE_Economy.pdf.

27 – "Size of the global Internet of Things (IoT) market from 2009 to 2019 (in billion U.S. dollars)," Statista — The Statistics Portal, accessed May 12, 2018, https://www.statista.com/statistics/485136/global-internet-of-things-market-size/.

28 – Brynjolfsson and McAfee, 251.

Chapter 11

1 – Digital Systems Research Center, "In Memoriam: J. C. R. Licklider," Stanford University, August 7, 1990, 1, https://web.stanford.edu/dept/SUL/library/extra4/sloan/mousesite/Secondary/Licklider.pdf.

2 – MarcelVEVO, "The Mother of All Demos, presented by Douglas Engelbart (1968)," video, 1:40:52, July 9, 2012, https://www.youtube.com/watch?v=yJDv-zdhzMY.

3 – Digital Systems Research Center, "In Memoriam: J. C. R. Licklider," 26.

4 – Neil Havermale, "Google I/O 2018 Maps and Lens," video, 7:56, May 8, 2018, https://www.youtube.com/watch?v=82CGhuwfk1I.

5 – Vanessa Ho, "'Heritage activists' preserve global landmarks ruined in war, threatened by time," Microsoft, April 23, 2018, https://news.microsoft.com/transform/heritage-activists-preserve-global-landmarks-ruined-in-war-threatened-by-time/.

6 – "Model Information," Honda, accessed May 12, 2018, http://owners.honda.com/vehicles/information/2018/Accord-Sedan/features/Head-Up-Display.

7 – "The BMW Heads Up Display," Herb Chambers BMW, accessed May 12, 2018, https://www.herbchambersbmw.com/new-bmw-features-head-up-display.htm

8 – Colin Ryan, "10 New Cars With Head-Up Displays," Autobytel, accessed May 12, 2018, https://www.autobytel.com/car-buying-guides/features/10-new-cars-with-head-up-displays-130663/.

9 – "How It Works," Be My Eyes, accessed May 12, 2018, https://www.bemyeyes.com/.

10 – Ian Steadman, "IBM's Watson is better at diagnosing cancer than human doctors," *Wired*, February 11, 2013, http://www.wired.co.uk/article/ibm-watson-medical-doctor.

11 – Pranav Rajpurkar et al., "Radiologist-Level Pneumonia Detection on Chest X-Rays with Deep Learning," December 25, 2017, https://arxiv.

org/pdf/1711.05225.pdf.

12 – Steadman, "IBM's Watson is better at diagnosing cancer than human doctors."

13 – Yun Liu et al., "Detecting Cancer Metastases on Gigapixel Pathology Images," March 8, 2017, https://arxiv.org/abs/1703.02442.

14 – Matthew Hudson, "Self-taught artificial intelligence beats doctors at predicting heart attacks," *Science*, April 14, 2017, http://www.science-mag.org/news/2017/04/self-taught-artificial-intelligence-beats-doctors-predicting-heart-attacks.

15 – "Vitality," NantHealth, accessed May 12, 2018, https://nanthealth.com/vitality/.

16 – "Medication Dispensers," MedReady Inc., accessed May 12, 2018, https://www.medreadyinc.net/.

17 – "How can technology make people in the world safer?," Jigsaw, accessed May 12, 2018, https://jigsaw.google.com/vision/.

18 – "About Us," Revolar, accessed May 12, 2018, https://revolar.com/pages/about-us.

19 – "About," Companion, accessed May 12, 2018, https://companion-app.io/.

20 – "About Us," Watch Over Me, accessed May 12, 2018, http://watchovermeapp.com/about/.

21 – Nicky Case, "How To Become A Centaur," *Science*, February 6, 2018, https://jods.mitpress.mit.edu/pub/issue3-case.

22 – Tercek, 222.

23 – "Alibaba's Jack Ma Sees Pain as Internet Disrupts Economy," *Bloomberg*, April 25, 2017, https://www.bloomberg.com/news/videos/2017-04-24/asia-savers-aren-t-investing-according-to-report-video.

Chapter 12

1 – "Home," Soul Machines, accessed May 12, 2018, https://www.soulmachines.com/.

2 – Rachel Metz, "Controlling VR with Your Mind," *MIT Technology Review*, March 22, 2017, https://www.technologyreview.com/s/603896/controlling-vr-with-your-mind/.

3 – Jacob Templin, "The US government just gave someone a $120-million robotic arm to use for a year," Quartz, February 2, 2018, https://qz.com/1194939/the-us-government-just-gave-someone-a-120-million-robotic-arm-to-use-for-a-year/.

4 – Steven Ashley, "Robotic Exoskeletons Are Changing Lives in Surprising Ways," NBC News, February 21, 2017, https://www.nbcnews.com/mach/innovation/robotic-exoskeletons-are-changing-lives-surprising-ways-n722676.

5 – "Request For Concept Papers On Exoskeleton Technologies For The Warfighter," Federal Business Opportunities, accessed May 12, 2018, https://www.fbo.gov/index?s=opportunity&mode=form&tab=core&id=9ba5618073c7bfa2d4abd42e1f5c4ee4.

6 – "Average Annual Miles per Driver by Age Group," U.S. Department of Transportation Federal Highway Administration, last modified March 30, 2018, https://www.fhwa.dot.gov/ohim/onh00/bar8.htm.

7 – Fred Lambert, "Tesla's fleet has driven 7.2 billion miles and its energy products produced 10.3 billion kWh," Electrek, accessed April 22, 2018, https://electrek.co/2018/04/22/tesla-fleet-miles-energy-products/.

8 – Elon Musk, "Master Plan, Part Deux," Tesla, July 20, 2016, https://www.tesla.com/blog/master-plan-part-deux.

9 – Dan Fagella, "Self-driving car timeline for 11 top automakers," *VentureBeat*, June 4, 2017, https://venturebeat.com/2017/06/04/self-driving-car-timeline-for-11-top-automakers/.

10 – Just Seven, "Tesla Autopilot Predicts Crash Compilation 2," video, 1:48, February 19, 2017, https://www.youtube.com/watch?v=--xITOqIB-CM.

11 – Michele Bertoncello and Dominik Wee, "Ten ways autonomous driving could redefine the automotive world," McKinsey Institute, June 2015, https://www.mckinsey.com/industries/automotive-and-assembly/our-insights/ten-ways-autonomous-driving-could-redefine-the-automotive-world.

12 – Andrew R. Long, "Urban Parking as Economic Solution," International Parking Institute, December 2013, https://www.parking.org/wp-content/uploads/2016/01/TPP-2013-12-Urban-Parking-as-Economic-Solution.pdf.

13 – Tim Henderson, "Why many teens don't want to get a driver's license," PBS, March 6, 2017, https://www.pbs.org/newshour/nation/many-teens-dont-want-get-drivers-license.

14 – Twenge, *iGen*, 26.

Chapter 13

1 – Arthur C. Clarke, *Profiles of the Future: An Enquiry into the Limits of the Possible* (New York: Henry Holt, 1984), 21.

2 – "Building" Bullitt Center, accessed May 28, 2018, http://www.bullittcenter.org/building/.

3 – Tercek, 151–2.

4 – David Rose, *Enchanted Objects: Design, Human Desire, and The Internet of Things* (New York: Scribner, 2014), 246-247.

5 – Cesar Harada, "A novel idea for cleaning up oil spills," filmed June 2012 at TEDxSummit, video, 14:23, https://www.ted.com/talks/cesar_harada_a_novel_idea_for_cleaning_up_oil_spills.

6 – "About," Made In Space, accessed May 12, 2018, http://madeinspace.us/mission/.

7 – Jalila Essaïdi, "A Wireless Station Within Reach of Everybody," Jalila Essaïdi, accessed May 12, 2018, http://jalilaessaidi.com/living-network/.

Chapter 14

1 – "The Antitrust Laws," Federal Trade Commission, accessed March 10, 2018, https://www.ftc.gov/tips-advice/competition-guidance/guide-antitrust-laws/antitrust-laws.

2 – Wu, 312.

3 – Federal Trade Commission, "The Antitrust Laws."

4 – "Mergers," Federal Trade Commission, accessed March 10, 2018,

https://www.ftc.gov/tips-advice/competition-guidance/guide-anti-trust-laws/mergers.

5 – "Council Regulation (EU) 2016/679 of 27 April 2016 on the protection of natural persons with regard to the processing of personal data and on the free movement of such data, and repealing Directive 95/46/EC (General Data Protection Regulation)," Official Journal of the European Union, L119, May 4, 2016, 1–88, https://eur-lex.europa.eu/legal-content/EN/TXT/PDF/?uri=CELEX:32016R0679&from=EN.

6 – "Robert Pepper; Stephen H. Schwartz; Edward W. Hayter; Eric Terrell v Apple, Inc.," United States Courts, accessed May 10, 2018, https://cdn.ca9.uscourts.gov/datastore/opinions/2017/01/12/14-15000.pdf .

7 – Raff and Raff, "The European Commission's Google Search Case: A Timeline of Significant Events."

8 – "Comments of the American Bar Association's Sections of Antitrust Law and International Law on the European Commission's Public Inception Impact Assessment on Fairness in Platform-to-Business Relations," The American Bar Association, January 9, 2018, 3, https://www.americanbar.org/content/dam/aba/administrative/antitrust_law/at_comments_20180109.authcheckdam.pdf.

9 – Stock quotes: "Alphabet Inc Class A," NASDAQ:GOOGL, Google Finance, July 11, 2018, retrieved from https://www.google.com/search?q=NASDAQ:GOOGL&e=4112296&tbm=fin&biw=1798&bih=903#scso=uid_1TBJW62dE8u4sQWJtI2oBQ_5:0; "Amazon.com, Inc.," ETR:AMZ, Google Finance, July 11, 2018, retrieved from https://www.google.com/search?q=NASDAQ:AMZ&safe=off&source=l-nms&tbm=fin&sa=X&ved=0ahUKEwil-Zmhj-LcAhUMybwKHfs2Ab-kQ_AUICigB&biw=1404&bih=664&dpr=2#scso=uid_2lNtW8a8OY-Lu8wWyw7yoAg_5:0; "Facebook, Inc. Common Stock" NASDAQ:FB, Google Finance, July 11, 2018, retrieved from https://www.google.com/search?biw=1798&bih=903&tbm=fin&ei=RDFJW_WIM8T4tAWDm-JSQAQ&q=facebook+stock&oq=facebook+stock&gs_l=finance-immersive.3..81i8k1l3.3376.4309.0.4798.8.8.0.0.0.0.135.712.5j3.8.0..2..0...1.1.64.finance-immersive..2.2.213....0.wJ9Js-_PkZY#scso=uid_SjFJW9PDLM-mYtgWLyY6AAw_5:0,uid_P1RtW9ayO8aF8wXi9rfgBg_5:0; "Apple Inc.," NASDAQ:AAPL, Google Finance, July 11, 2018, retrieved from https://www.google.com/search?biw=1798&bih=903&tbm=fin&ei=S-jFJW9PDLMmYtgWLyY6AAw&q=apple+stock&oq=apple+stock&gs_l=-finance-immersive.3...35411.37559.0.37653.11.9.2.0.0.0.101.650.7j1.8.0..2..0...1.1.64.finance-immersive..2.5.268...81j81i8k1.0.-qkSF-dHL-NU#scso=uid_cTFJW9XOO82OsQW-jI7ABA_5:0,uid_g1RtW5e-3B4yy8QWj84nIAg_5:0; "Dow Jones Industrial Average," INDEXDJX:.DJI, Google Finance, July 11, 2018, retrieved from https://www.google.com/search?biw=1798&bih=903&tbm=fin&ei=cTFJW9XOO82OsQW-jI-7ABA&q=djia&oq=djia&gs_l=finance-immersive.3..81l3.61658.62277.0.62476.4.4.0.0.0.0.135.374.3j1.4.0..2..0...1.1.64.finance-immersive..0.4.372....0.eASX1VD-hOw#scso=uid_sTFJW8-XIYzwsQXC7re4D-Q_5:0,uid_slRtW6uZPITO8wXQgJKIBA_5:0.

10 – The American Bar Association, "Comments of the American Bar Association's Sections of Antitrust Law and International Law on the European Commission's Public Inception Impact Assessement on

Fairness in Platform-to-Business Relations," 3.

11 – "Most famous social network sites worldwide as of April 2018, ranked by number of active users (in millions)," Statista — The Statistics Portal, accessed March 13, 2018, https://www.statista.com/statistics/272014/global-social-networks-ranked-by-number-of-users/.

12 – "Google Home Is 6 Times More Likely to Answer Your Question Than Amazon Alexa," *Adweek*, June 23, 2017, http://www.adweek.com/digital/google-home-is-6-times-more-likely-to-answer-your-question-than-amazon-alexa/.

13 – Christopher Heine, "Search Engine Market Share," Net Marketshare, accessed March 13, 2018, https://www.net-marketshare.com/search-engine-market-share.aspx?options=%7B%22filter%22%3A%7B%22%24and%22%3A%5B%7B%22deviceType%22%3A%7B%22%24in%22%3A%5B%22Desktop%2Flaptop%22%5D%7D%7D%5D%7D%2C%22dateLabel%22%3A%22Trend%22%2C%22attributes%22%3A%22share%22%2C%22group%22%3A%22searchEngine%22%2C%22sort%22%3A%7B%22share%22%3A-1%7D%2C%22id%22%3A%22searchEnginesDesktop%22%2C%22dateInterval%22%3A%22Monthly%22%2C%22dateStart%22%3A%222017-02%22%2C%22dateEnd%22%3A%222018-01%22%2C%22segments%22%3A%22-1000%22%7D.

14 – "Antitrust: Commission sends Statement of Objections to Google on comparison shopping service," European Commission, April 15, 2015, http://europa.eu/rapid/press-release_MEMO-15-4781_en.htm.

15 – Silicon Valley Bank, U.S. Startup Outlook 2017, February 17, 2017, 9, https://www.slideshare.net/SVBFinancial/us-startup-outlook-report-2017-72285360 (accessed July 11, 2018).

16 – Craig Doidge, G. Andrew Karolyi, and René M. Stulz, "The US Listing Gap," (working paper 21181, National Bureau of Economic Research, Cambridge, MA, May 2015), 9, http://www.nber.org/papers/w21181.pdf.

17 – Spectrum Sports, Inc., et al. v. McQuillan et vir, dba Sorboturf Enterprises, 506 U.S. 447 (9th Cir. 1993), 12, https://supreme.justia.com/cases/federal/us/506/447/case.pdf.

Chapter 15

1 – "News Use Across Social Media Platforms 2017," Pew Research Center, September 7, 2017, http://www.journalism.org/2017/09/07/news-use-across-social-media-platforms-2017/.

2 – "People, Power and Technology: The 2018 Digital Understanding Report," doteveryone, 2018, http://understanding.doteveryone.org.uk/files/Doteveryone_PeoplePowerTechDigitalUnderstanding2018.pdf.

3 – Laura Locke, "The Future of Facebook," *Time*, July 17, 2017, http://content.time.com/time/business/article/0,8599,1644040,00.html.

4 – Stanford Graduate School of Business, "Chamath Palihapitiya, Founder and CEO Social Capital, on Money as an Instrument of Change.

5 – Emerging Technology from the arXiv, "First Evidence That Social

Bots Play a Major Role in Spreading Fake News," *MIT Technology Review*, August 7, 2017, https://www.technologyreview.com/s/608561/ first-evidence-that-social-bots-play-a-major-role-in-spreading-fake-news/.

6 – Nicholas Confessore et al., "The Follow Factory," *New York Times*, January 27, 2018, https://www.nytimes.com/interactive/2018/01/27/ technology/social-media-bots.html.

7 – Jim Zarroli, "Can't Buy A Ticket To That Concert You Want To See? Blame Bots," NPR, January 28, 2016, https://www.npr.org/sections/thetwo-way/2016/01/28/464708137/cant-buy-a-ticket-to-that-concert-you-want-to-see-blame-bots.

8 – Samantha Raphelson, "'Grinch Bots' Attempt To Steal Christmas By Driving Up Toy Prices," NPR, December 25, 2017, https://www.npr. org/2017/12/05/568624246/grinch-bots-attempt-to-steal-christmas-by-driving-up-toy-prices.

9 – Confessore et al., "The Follow Factory."

10 – Igal Zeifman, "'Bot Traffic Report 2016," Incapsula, January 24, 2017, https://www.incapsula.com/blog/bot-traffic-report-2016.html.

11 – Dr. Angelos Keromytis, "Enhanced Attribution," Defense Advanced Research Project Agency, accessed March 13, 2018, https://www.darpa. mil/program/enhanced-attribution.

12 – Cory Doctrow, "Chaos Computer Club claims it can unlock Iphones with fake fingers/cloned fingerprints," BoingBoing, September 22, 2013, https://boingboing.net/2013/09/22/chaos-computer-club-claims-it.html.

13 – Jan Krissler, "Jan Krissler World Famous Hackers Who Hack Fingerprint," Jan Krissler, August 19, 2016, http://jankrissler.blogspot. com/2016/08/jan-krissler-world-famous-hackers-who.html.

14 – 46halbe, "Chaos Computer Clubs breaks iris recognition system of the Samsung Galaxy S8," Chaos Computer Club, May 22, 2017, https://www.ccc.de/en/updates/2017/iriden.

15 – "About," what3words, accessed March 13, 2018, https://what-3words.com/about/.

16 – "Definitive data and analysis for the mobile industry," GSMA Intelligence, accessed March 13, 2018, https://www.gsmaintelligence.com/.

17 – Dr. Jennifer Roberts, "Cyber-Hunting at Scale (CHASE)," Defense Advanced Research Projects Agency, accessed March 13, 2018, https:// www.darpa.mil/program/cyber-hunting-at-scale.

18 – Austin Carr and Harry McCracken, "'Did We Create This Monster?' How Twitter Turned Toxic," *Fast Company*, April 4, 2018, https://www.fastcompany.com/40547818/did-we-create-this-monster-how-twitter-turned-toxic.

19 – "New York City, New York Population 2018," World Population Review, accessed April 6, 2018, http://worldpopulationreview.com/us-cities/new-york-city-population/.

20 – Joseph Marks, "DARPA Wants to Merge Human and Computer Cyber Defenders," *Defense One*, April 23, 2018, https://www.defense-one.com/technology/2018/04/darpa-wants-merge-human-and-computer-cyber-defenders/147652/.

21 – "Applying Computer-Human Collaboration to Accelerate Detec-

tion of Zero-Day Vulnerabilities," Defense Advanced Research Projects Agency, April 18, 2018, https://www.darpa.mil/news-events/2018-04-18a.

Chapter 16

1 – "Planning Outline for the Construction of a Social Credit System (2014–2020)," China Copyright Media, accessed April 25, 2015, https://chinacopyrightandmedia.wordpress.com/2014/06/14/planning-out-line-for-the-construction-of-a-social-credit-system-2014-2020/.
2 – Rachel Botsman, "Big data meets Big Brother as China moves to rate its citizens," *Wired*, October 21, 2017, http://www.wired.co.uk/arti-cle/chinese-government-social-credit-score-privacy-invasion.
3 – Botsman, "Big data meets Big Brother as China moves to rate its citizens."
4 – James Vincent, "Twitter taught Microsoft's AI chatbot to be a racist asshole in less than a day," *The Verge*, March 24, 2016, https://www.theverge.com/2016/3/24/11297050/tay-microsoft-chatbot-racist.
5 – Tom Simonite, "When it comes to gorillas, Google Photos remains blind," *Wired*, January 11, 2018, https://www.wired.com/story/when-it-comes-to-gorillas-google-photos-remains-blind/.
6 – Graeme McMillan, "It's Not You, It's It: Voice Recognition Doesn't Recognize Women," *Time*, June 1, 2011, http://techland.time.com/2011/06/01/its-not-you-its-it-voice-recognition-doesnt-recognize-women/.
7 – Mariya Yao, "Chihuahua or muffin? My search for the best computer vision API," freeCodeCamp, October 12, 2017, https://medium.freeco-decamp.org/chihuahua-or-muffin-my-search-for-the-best-computer-vision-api-cbda4d6b425d.
8 – Friedrich Hayek, *Individualism and Economic Order* (Chicago: University of Chicago Press, 1948), 77.
9 – "About," Data Does Good, accessed February 27, 2018, https://m.datadoesgood.com/?section=mission.
10 – Christina Aperjis and Bernardo Huberman, "A market for ubiased private data: Paying individuals according to their privacy attitudes," First Monday, May 7, 2012, http://firstmonday.org/ojs/index.php/fm/arti-cle/view/4013/3209.
11 – Harari, 132.
12 – U.S. Army, "Robotic and Autonomous Systems Strategy," United States Army Training and Doctrine Command, March, 2017. http://www.tradoc.army.mil/FrontPageContent/Docs/RAS_Strategy.pdf.
13 – Janosch Deckler, "Attack on the killer robots," *Politico*, February 24, 2018, https://www.politico.eu/article/attack-killer-robots-autono-mous-weapons-drones/.
14 – Page and Brin, "The Anatomy of a Large-Scale Hypertextual Search Engine."
15 – "Services," ORCAA, accessed March 23, 2018, http://www.oneilrisk.com/.
16 – "Council Regulation (EU) 2016/679 of 27 April 2016 on the pro-tection of natural persons with regard to the processing of personal

data and on the free movement of such data, and repealing Directive 95/46/EC (General Data Protection Regulation)."

Chapter 17

1 – "18 U.S. Code § 1708 – Theft or receipt of stolen mail matter generally," Legal Information Institute, Cornell Law School, accessed March 23, 2018, https://www.law.cornell.edu/uscode/text/18/1708.

2 – Kenneth A. Bamberger and Deirdre K. Mulligan, *Privacy on the Ground: Driving Corporate Behavior in the United States and Europe* (Cambridge, MA: The MIT Press, 2015), 47–49.

3 – Nick Bilton. "Steve Jobs Was a Low-Tech Parent," *New York Times*, May 24, 2016, https://www.nytimes.com/2014/09/11/fashion/steve-jobs-apple-was-a-low-tech-parent.html?mtrref=www.google.com.

4 – Denver Nicks. "Mark Zuckerberg Bought Four Houses Just to Tear Them Down," Time, accessed March 23, 2018, http://time.com/money/4346766/mark-zuckerberg-houses/.

5 – Paul Lewis, "'Our minds can be hijacked': the tech insiders who fear a smartphone dystopia," *Guardian*, October 6, 2017, https://www.theguardian.com/technology/2017/oct/05/smartphone-addiction-silicon-valley-dystopia.

6 – Mary Madden and Lee Rainie, "Americans' Attitudes About Privacy, Security and Surveillance," Pew Research Center, May 20, 2015, http://www.pewinternet.org/2015/05/20/americans-attitudes-about-privacy-security-and-surveillance/.

7 – Lee Rainie et al., "Anonymity, Privacy, and Security Online," Pew Research Center, September 5, 2013, http://www.pewinternet.org/2013/09/05/anonymity-privacy-and-security-online/.

8 – doteveryone, "People, Power and Technology: The 2018 Digital Attitudes Report," 2018, 18, http://attitudes.doteveryone.org.uk/files/People%20Power%20and%20Technology%20Doteveryone%20Digital%20Attitudes%20Report%202018.pdf.

9 – Rainie et al., "Anonymity, Privacy, and Security Online."

10 – Madden and Rainie, "Americans' Attitudes About Privacy, Security and Surveillance."

11 – The White House, "Consumer Data Privacy in a Networked World: A Framework for Protecting Privacy and Promoting Innovation in the Global Digital Economy," February 2012, doi: 10.29012/jpc.v4i2.623.

12 – Alex Pentland, *Social Physics: How Social Networks Can Make Us Smarter* (New York: Penguin Books, 2015), 177.

13 – "CAN-SPAM Act: A Compliance Guide for Business," Federal Trade Commission, September 2009, https://www.ftc.gov/tips-advice/business-center/guidance/can-spam-act-compliance-guide-business.

14 – Aleecia McDonald and Lorrie Cranor, *I/S: A Journal of Law and Policy for the Information Society*, vol. 4, no. 3 (2008), 543–68, http://hdl.handle.net/1811/72839.

15 – Laura Brandimarte, Alessandro Acquisti, and George Loewenstein, "Misplaced Confidences: Privacy and the Control Paradox," *Social Psychological and Personality Science* 4, no. 3, (May 2013): 340–47, https://doi.org/10.1177/1948550612455931.

16 – Alessandro Acquisti, "Privacy in Electronic Commerce and the Economics of Immediate Gratification," in *EC '04 Proceedings of the 5th ACM Conference on Electronic Commerce* (2004): 21–29, doi: 10.1145/988772.988777.

17 – Paul A. Pavlou, Huigang Liang, and Yajiong Xue, "Understanding and Mitigating Uncertainty in Online Environments: A Principal-Agent Perspective," *MIS Quarterly* 31, no. 1 (January 2006): 105–36, https://ssrn.com/abstract=2380164.

18 – "Regulating for Results: Strategies and Priorities for Leadership and Engagement," (discussion paper, Center for Information Policy Leadership, Hunton & Williams LLC, September 25, 2017), 20, https://www.informationpolicycentre.com/uploads/5/7/1/0/57104281/cipl_final_draft_regulating_for_results_-_strategies_and_priorities_for_leadership_and_engagement.pdf.

19 – "Privacy Impact Assessments," Federal Trade Commission, accessed March 23, 2018, https://www.ftc.gov/site-information/privacy-policy/privacy-impact-assessments

20 – Bamberger and Mulligan, 95.

21 – "Regulating for Results: Strategies and Priorities for Leadership and Engagement," 16.

22 – "Regulating for Results: Strategies and Priorities for Leadership and Engagement," 19.

23 – Tom Warren, "Google fined a record $5 billion by the EU for Android antitrust violations," *The Verge*, July 18, 2018, https://www.theverge.com/2018/7/18/17580694/google-android-eu-fine-antitrust.

Chapter 18

1 – Miles Brundage et al., "The Malicious Use of Artificial Intelligence: Forecasting, Prevention, and Mitigation," February, 2018, 13, https://arxiv.org/pdf/1802.07228.pdf.

2 – BostonDynamics, "What's new, Atlas?," YouTube video, November 16, 2017, 0:54, https://www.youtube.com/watch?v=fRj34o4hN4I.

3 – Klint Finley, "AI Fighter Pilot Beats A Human, But No Need To Panic (Really)," *Wired*, June 29, 2016, https://www.wired.com/2016/06/ai-fighter-pilot-beats-human-no-need-panic-really/.

4 – Miles Brundage et al., "The Malicious Use of Artificial Intelligence: Forecasting, Prevention, and Mitigation," 32.

5 – Greg Allen and Taniel Chen, "Artificial Intelligence and National Security," Harvard Kennedy School Belfer Center for Science and International Affairs, July 2017, 58–69, https://www.belfercenter.org/sites/default/files/files/publication/AI%20NatSec%20-%20final.pdf.

6 – Wu, 316.

7 – "Trends in Current Cigarette Smoking Among High School Students and Adults, United States, 1965–2014," Centers for Disease Control and Prevention, March 3, 2016, https://www.cdc.gov/tobacco/data_statistics/tables/trends/cig_smoking/index.htm.

8 – Aoibhinn McBride, "Netflix user who binge-watched The Office solidly for a week during bout of depression reveals site emailed them to see if they were OKAY in latest abuse of viewer data," *Daily Mail*,

December 14, 2017, http://www.dailymail.co.uk/femail/article-5172881/
Netflix-user-binge-watched-Office-non-stop-week.html.
9 – "New Letter from Jana Partners and Calstrs to Apple Inc.," Jana
Partners LLC, June 4, 2018 https://thinkdifferentlyaboutkids.com/.

Chapter 19

1 – Carl Benedikt Frey and Michael A. Osborne, "The Future of Employment: How Susceptible Are Jobs to Computerisation?," (working
paper, Oxford Martin School, University of Oxford, September 2013), 1,
http://www.oxfordmartin.ox.ac.uk/downloads/academic/The_Future_
of_Employment.pdf.
2 – James Manyika et al., "Jobs lost, jobs gained: What the future of
work will mean for jobs, skills, and wages," McKinsey Institute, November 2017, https://www.mckinsey.com/featured-insights/future-of-or-
ganizations-and-work/jobs-lost-jobs-gained-what-the-future-of-work-
will-mean-for-jobs-skills-and-wages.
3 – Ljubica Nedelkoska and Glenda Quintini, "Automation, skills use
and training," *OECD Social, Employment and Migration Working Papers*, no. 202 (March 2018), https://doi.org/10.1787/2e2f4eea-en.
4 – Ginni Rometty, "Ginni Rometty on How AI Is Going to Transform
Jobs—All of Them," *Wall Street Journal*, January 17, 2018 https://www.
wsj.com/articles/ginni-rometty-on-how-ai-is-going-to-transform-jobsa-
ll-of-them-1516201040.
5 – Tercek, 237.
6 – Sir Ken Robinson, *Creative Schools: The Grassroots Revolution
That's Transforming Education* (New York: Penguin Books, 2015), 20.
7 – Henry M. Levin and Cecilia E. Rouse. "The True Cost of High School
Dropouts," *New York Times*, January 25, 2012, http://www.nytimes.
com/2012/01/26/opinion/the-true-cost-of-high-school-dropouts.html.
8 – "A Nation At Risk," U.S. Department of Education, April 1983,
https://www2.ed.gov/pubs/NatAtRisk/risk.html.
9 – David Kastberg, Jessica Ying Chan, and Gordon Murray, "Performance of U.S. 15-Year-Old Students in Science, Reading, and Mathematics Literacy in an International Context: First Look at PISA 2015"
(Washington, DC: National Center for Education Statistics, U.S. Department of Education, December 2016), 17, 20, 23, https://nces.ed.gov/
pubs2017/2017048.pdf.
10 – Michelle Jamrisko & Wei Lu, "The U.S. Drops Out of the Top 10
in Innovation Ranking," *Bloomberg*, January 22, 2018 https://www.
bloomberg.com/news/articles/2018-01-22/south-korea-tops-global-in-
novation-ranking-again-as-u-s-falls.
11 – Andy McCreedy, "The 'Creativity Problem' and the Future of the
Japanese Workforce," Asia Program Special Report, no. 121 (June
2004): 2, 3, https://www.wilsoncenter.org/sites/default/files/asiarpt121.
pdf.
12 – Kastberg, Chan, and Murray.
13 – P. Sahlberg. *Finnish Lessons 2.0: What Can the World Learn from
Educational Change in Finland?* (New York: Teachers College Press,
2014).

14 – Robinson, 60–61.
15 – Jamrisko and Lu, "The U.S. Drops Out of the Top 10 in Innovation Ranking."
16 – Kastberg, Chan, and Murray.
17 – "Policies & Programs," Korean Ministry of Education, accessed January 27, 2018, http://english.moe.go.kr/sub/info.do?m=040101&s=english.
18 – Melissa Harrell and Lauren Barbato, "Great managers still matter: the evolution of Google's Project Oxygen," re:Work, February 27, 2018, https://rework.withgoogle.com/blog/the-evolution-of-project-oxygen/.
19 – Valerie Strauss, "The surprising thing Google learned about its employees — and what it means for today's students," *Washington Post*, December 20, 2017, https://www.washingtonpost.com/news/answer-sheet/wp/2017/12/20/the-surprising-thing-google-learned-about-its-employees-and-what-it-means-for-todays-students/?utm_term=.b5ca259d4f06.
20 – IBM, "The Enterprise of the Future: IBM Global CEO Study," 2008, https://www-935.ibm.com/services/uk/gbs/pdf/ibm_ceo_study_2008.pdf.
21 – "The World Fact Book," Central Intelligence Agency, accessed March 23, 2018 https://www.cia.gov/library/publications/the-world-fact-book/fields/2103.html.
22 – Miller, Coldicutt, and Kitcher, "People, Power and Technology: The 2018 Digital Understanding Report," 4.
23 – Robinson, 154.
24 – Robinson, 52.
25 – Frey and Osborne, "The Future of Employment: How Susceptible Are Jobs to Computerisation?," 57–72.

Chapter 20

1 – "Wolf Restoration," National Park Service, accessed March 23, 2018 https://www.nps.gov/yell/learn/nature/wolf-restoration.htm.
2 – "1995 Reintroduction of Wolves in Yellowstone," Yellowstone National Park, accessed March 23, 2018, https://www.yellowstonepark.com/park/yellowstone-wolves-reintroduction.
3 – "Wolf Reintroduction Changes Ecosystem," Yellowstone National Park, accessed March 23, 2018, https://www.yellowstonepark.com/things-to-do/wolf-reintroduction-changes-ecosystem.
4 – "Women in Computer Science: Getting Involved in STEM," ComputerScience.org, accessed March 23, 2018, https://www.computerscience.org/resources/women-in-computer-science/.
5 – Kieran Snyder, "Why women leave tech: It's the culture, not because 'math is hard'," *Fortune*, October 24, 2014, http://fortune.com/2014/10/02/women-leave-tech-culture/.
6 – Nadya A. Fouad, "Leaning in, but Getting Pushed Back (and Out)," (PowerPoint presentation, American Psychological Association Annual Convention, Washington, D.C., August 7–10, 2014), http:// www.apa.org/news/press/releases/2014/08/pushed-back.pdf.
7 – Alison Wood Brooks et al., "Investors prefer entrepreneurial ventures

pitched by attractive men," *Proceedings of the National Academy of Sciences* 111, no. 12 (March 2014): 4427–31, doi: 10.1073/pnas.1321202111.

8 – Hannah Riley Bowles, Linda Babcock, and Lei Lai, "It Depends Who is Asking and Who You Ask: Social Incentives for Sex Differences in the Propensity to Initiate Negotiation," (KSG Working Paper No. RWP05-045, July 2005), doi: 10.2139/ssrn.779506

9 – Ernesto Reuben, Paola Sapienza, and Luigi Zingales, "How stereo-types impair women's careers in science," *Proceedings of the National Academy of Sciences* 111, no. 12 (March 2014): 4403–08, doi: 10.1073/pnas.1314788111.

10 – "Median annual earnings by sex, race and Hispanic ethnicity March 1960-2016," United States Department of Labor, accessed April 6, 2018, https://www.dol.gov/wb/stats/NEWSTATS/facts/earn_earnings_ratio.htm#one; "Percentage of the U.S. population who have complet-ed four years of college or more from 1940 to 2017, by gender," Statista – The Statistics Portal, accessed April 6, 2018, https://www.statista.com/statistics/184272/educational-attainment-of-college-diploma-or-high-er-by-gender/.

11 – ComputerScience.org, "Women in Computer Science: Getting Involved in STEM."

12 – Buck Gee and Denise Peck, "The Illusion of Asian Success: Scant Progress for Minorities Cracking the Glass Ceiling from 2007-2015," Ascend Pan-Asian Leaders, 2017, https://c.ymcdn.com/sites/www.ascendleadership.org/resource/resmgr/research/TheIllusionofAsian-Success.pdf.

13 – Christopher Woods, "Bilingualism, Scribal Learning, and the Death of Sumerian," *Margins of Writing, Origins of Culture*, ed. Seth L. Sand-ers (USA: The University of Chicago, 2006): 95–124, https://oi.uchicago.edu/sites/oi.uchicago.edu/files/uploads/shared/docs/ois2_2007.pdf.

14 – "Hindu-Arabic numerals," Encyclopædia Britannica, accessed April 6, 2018, https://www.britannica.com/topic/Hindu-Arabic-numerals.

15 – "Foundations of Physics: Thales and Euclid," Blogs at UMass Amherst, accessed April 6, 2018, https://blogs.umass.edu/p139ell/2012/09/30/foundations-of-physics-thales-and-euclid/.

16 – Eric Weiner, *The Geography of Genius: Lessons from the World's Most Creative Places* (New York: Simon & Schuster, 2016), 264.

17 – Stuart Anderson, "Immigrants and Billion Dollar Startups," (Na-tional Foundation for American Policy, NFAP Policy Brief, March 2006), 1, https://nfap.com/wp-content/uploads/2016/03/Immigrants-and-Bil-lion-Dollar-Startups.NFAP-Policy-Brief.March-2016.pdf.

18 – Vala Afshar (@ValaAfshar), Tweet, May 18, 2018 (11:22 p.m.), https://twitter.com/ ValaAfshar/status/997693832828350464.

19 – Richard Florida, "Immigrants Boost the Wages, Income and Eco-nomic Output of Cities," *City Lab*, April 25, 2013, https://www.citylab.com/equity/2013/04/how-immigration-helps-cities/5323/.

20 – Chang-Tai Hsieh et al., "The Allocation of Talent and U.S. Econom-ic Growth," (NBER working paper no. 18693, January 2013), http://www.nber.org/papers/w18693.

21 – James Damore, "Google's Ideological Echo Chamber: How bias clouds our thinking about diversity and inclusion," July 2017, https://assets.document-

cloud.org/documents/3914586/Googles-Ideological-Echo-Chamber.pdf.

Chapter 21

1 – Eric Wasatonic, "(#0119) HP Origins," copied from an HP Promotional DVD 2005, video, 25:46, October 30, 2013, https://www.youtube.com/watch?v=hLORM1TcE1A&t=.
2 – theinnovationchannel, "Steve Jobs - Disruptive Innovation Documentary - One Last thing (HD - Full Length)," video, 55:55, January 6, 2012, https://www.youtube.com/watch?v=pnoeSvHAJ9I.
3 – Bloomberg, "Mark Zuckerberg: Building the Facebook Empire," video, 45:17, May 28, 2013, https://www.youtube.com/watch?v=5WiDI-hIkPoM.
4 – Gloria G. Guzman, "Household Income: 2016," United States Census Bureau, September 2017, https://www.census.gov/content/dam/Census/library/publications/2017/acs/acsbr16-02.pdf.
5 – Wendy Lee, "Why tech companies spend so much on holiday parties," *SFGate*, December 14, 2014, https://www.sfgate.com/business/article/Why-tech-companies-spend-so-much-on-holiday-5954418.php.
6 – Sangeeta Singh-Kurtz, "Silicon Valley firms are hiring models to ogle at this year's holiday parties," Quartz, December 9, 2017, https://work.qz.com/1151342/silicon-valley-firms-are-hiring-models-to-ogle-at-this-years-holiday-parties/.
7 – Adam Smith, *The Wealth of Nations* (New York: Borzoi, 1999), 451.
8 – "Product," SALt, accessed March 23, 2018, http://saltph.strikingly.com/#product.
9 – "Bio-Cassava Bag," Avani, accessed June 2, 2018, https://www.avanieco.com/portfolio-item/bio-cassava-bag/.
10 – "Why Edible Cutlery," Bakeys, accessed June 2, 2018, http://www.bakeys.com/why-edible-cutlery/.
11 – "Allianz invests $100 mln in micro-insurer BIMA," *Reuters*, February 28, 2017, https://www.reuters.com/article/allianz-investment-bima/allianz-invests-100-mln-in-micro-insurer-bima-idUSL8N1OJ35Q.
12 – "What We Do," infarm, accessed June 2, 2018, https://infarm.de/#what-we-do.
13 – "Our Approach," Think Whole Person Healthcare, accessed May 29, 2018, https://thinkhealthcare.org/our-approach/.
14 – Gary Hamel and Michele Zanini, "A Few Unicorns Are No Substitute for a Competitive, Innovative Economy," Wired, accessed June 2, 2018, https://hbr.org/2017/02/a-few-unicorns-are-no-substitute-for-a-competitive-innovative-economy.
15 – Jordan Golson, "Tesla just gave all it's patents away to competitors," *Wired*, June 12, 2014, https://www.wired.com/2014/06/tesla-just-gave-all-its-patents-away-to-competitors/.
16 – Eric Reis, *The Startup Way: How Modern Companies Use Entrepreneurial Management to Transform Culture and Drive Long-Term Growth* (New York: Currency, 2017), 325.
17 – ExpovistaTV, "Davos 2018: Jack Ma's Keys to Success: Technology, Women, Peace and Never Complain," video, 50:18, January 24, 2018, https://www.youtube.com/watch?v=-nSbkywGf-E.

Chapter 22

1 – Louis Hansen, "Bay Area homes deliver record-breaking re-turns," *Mercury News*, Februay 28, 2018, https://www.mercurynews.com/2018/02/28/bay-area-home-prices-continue-to-rise-in-record-breaking-streak/.

2 – Hannah Norman, "If you make $117,400 a year, you qualify for low-income housing in San Francisco, HUD says," *San Francisco Business News*, June 26, 2018, https://www.bizjournals.com/sanfrancisco/news/2018/06/25/sf-households-six-figures-low-income.html.

3 – World Bank Group, "World Development Report 2016: Digital Dividends," 2016, 3, http://documents.worldbank.org/curated/en/896971468194972881/pdf/102725-PUB-Replacement-PUBLIC.pdf.

4 – Janelle Jones and Ben Zipperer, "Unfulfilled Promises," Economic Policy Institute, January 5, 2018, https://www.epi.org/publication/unfulfilled-promises-amazon-warehouses-do-not-generate-broad-based-employment-growth/.

5 – Zach Schiller, "More Ohio Amazon workers relying on food aid," Policy Matters Ohio, accessed April 7, 2018, https://www.policymattersohio.org/files/news/1518amazonpr.pdf.

6 – Claire Brown, "Amazon gets tax breaks while its employees rely on food stamps, new data shows," Economic Policy Institute, April 19, 2018, https://theintercept.com/2018/04/19/amazon-snap-subsidies-warehousing-wages/.

7 – Theresa Agovino, "How much is a NYC taxi medallion worth these days?," CBS News, April 14, 2017, https://www.cbsnews.com/news/how-much-is-a-nyc-taxi-medallion-worth-these-days/.

8 – Ginia Bellafonte, "A Driver's Suicide Reveals the Dark Side of the Gig Economy," *New York Times*, February 6, 2018, https://www.nytimes.com/2018/02/06/nyregion/livery-driver-taxi-uber.html.

9 – Daron Acemoglu and Pascual Restrepo, "Artificial Intelligence, Automation, and the Economy," (NBER Working Paper No. 24196, Executive Office of the President of the United States, January 2018), 40, doi: 10.3386/w24196.

10 – Brandi Janssen, "Safety Watch: Suicide rate among farmers at historic high," *Iowa Farmer Today*, December 10, 2016, 2, http://www.public-health.uiowa.edu/gpcah/wp-content/uploads/2017/04/Safety-Watch_-Suicide-rate-among-farmers-at-historic-high.pdf.

11 – Galloway, 69.

12 – Matt Orfalea, "Elon Musk says Universal Basic Income is 'going to be necessary'," video, 1:56, February 16, 2017, https://www.youtube.com/watch?v=e6HPdNBicM8.

13 – Patrick Gillispie, "Mark Zuckerberg supports universal basic income. What is it?," CNN, May 26, 2017, http://money.cnn.com/2017/05/26/news/economy/mark-zuckerberg-universal-basic-income/index.html.

14 – "Plutonomy: Buying Luxury, Explaining Global Imbalances," Citi Group, October 16, 2005, 24–25, http://www.lust-for-life.org/Lust-For-Life/CitigroupImbalances_October2009/CitigroupImbalances_October2009.pdf.

15 – Ajay Tandon et al., "Measuring Overall Health System Performance

for 191 Countries," (GPE Discussion Paper Series: No. 30, World Health Organization, 2000), http://www.who.int/healthinfo/paper30.pdf.

16 – Monica Lungu, "Tuition-Free Universities in Finland, Norway and Germany in 2018," Study Portals, December 19, 2017, https://www.mastersportal.com/articles/1042/tuition-free-universities-in-finland-norway-and-germany-in-2018.html.

17 – Antti Jauhiainen and Joona-Hermanni Mäkinen, "Universal Basic Income Didn't Fail in Finland. Finland Failed It," *New York Times*, May 2, 2018, https://www.nytimes.com/2018/05/02/opinion/universal-basic-income-finland.html.

18 – "Italians demand M5S basic income – even though the party's not in power," The Local, March 9, 2018, https://www.thelocal.it/20180309/five-star-movement-universal-basic-income-italy-election.

19 – "The first study of basic income in the United States.," Y Combinator, accessed April 7, 2018, https://basicincome.ycr.org/.

20 – Steven Hill, New Economy, New Social Contract: A Plan for a Safety Net in a Multiemployer World (New America, 2005), 7, 8, https://na-production.s3.amazonaws.com/documents/New_Economy_Social_Contract.pdf.

21 – Sara Horowitz and Fabio Rosati, "53 Million Americans Are Freelancing, New Survey Finds," Freelancers Union, September 3, 2014, https://www.freelancersunion.org/blog/dispatches/2014/09/04/53million.

22 – "QuickBooks Survey Reveals New Era of Entrepreneurship," Intuit, August 13, 2015, https://investors.intuit.com/press-releases/press-release-details/2015/Intuit-Forecast-76-Million-People-in-On-Demand-Economy-by-2020/default.aspx.

23 – Syllabus, Epic Systems Corporation. v. Jacob Lewis, 584 U.S. ___ (2018) (no. 16–285), https://www.supremecourt.gov/opinions/17pdf/16-285_q8l1.pdf.

24 – "Equal treatment for all agency workers," European Commission, October 22, 2008, http://ec.europa.eu/social/main. jsp?catId=89&langId=en&newsId=410&furtherNews=yes; "Working Conditions," European Commission, accessed May 3, 2018, http://ec.europa.eu/social/main.jsp?catId=706&langId=en&intPageId=203; "Part-time work," Europa, July 4, 2006, http://europa.eu/legislation_summaries/employment_and_social_policy/employment_rights_and_work_organisation/c10416_ en.htm.

25 – Michael Grabell and Lena Groeger, "Temp Worker Regulations Around the World," ProPublica, February 24, 2014, http://projects.propublica.org/graphics/temps-around-the-world.

26 – Dynamex Operations West, Inc. v. Superior Court of Los Angeles County, Super. Ct. No. BC332016 (2018), http://www.courts.ca.gov/opinions/documents/S222732.PDF.

27 – "HB 2812," Washington State Legislature, accessed March 28, 2018, http://apps2.leg.wa.gov/billsummary?BillNumber=2812&Year=2017&BillNumber=2812&Year=2017.

28 – Deborah Wetzel and Valor Econômico, "Bolsa Família: Brazil's Quiet Revolution," The World Bank, November 4, 2013, http://www.worldbank.org/en/news/opinion/2013/11/04/bolsa-familia-Brazil-quiet-revolution.

29 – Micah Maidenberg and Aurelien Breeden, "Google Wins Tax Case in France, Avoiding $1.3 Billion Bill," *New York Times*, July 12, 2017, https://www.nytimes.com/2017/07/12/business/13google.html.

30 – Simon Bowers, "Leaked Documents Expose Secret Tale Of Apple's Offshore Island Hop," International Consortium of Investigative Journalists, November 6, 2017, https://www.icij.org/investigations/paradise-papers/apples-secret-offshore-island-hop-revealed-by-paradise-papers-leak-icij/.

31 – Nick Wingfield and Patricia Cohen, "Amazon Plans Second Headquarters, Opening a Bidding War Among Cities," *New York Times*, September 7, 2017, https://www.nytimes.com/2017/09/07/technology/amazon-headquarters-north-america.html.

32 – Kevin J. Delaney, "The robot that takes your job should pay taxes, says Bill Gates," Quartz, February 18, 2017, https://qz.com/911968/bill-gates-the-robot-that-takes-your-job-should-pay-taxes/.

33 – Daron Acemoglu and Pascual Restrepo, "Artificial Intelligence, Automation, and the Economy," 43.

34 –Standard & Poor's Ratings Services, "How Increasing Income Inequality Is Dampening U.S. Economic Growth, And Possible Ways To Change The Tide," August 5, 2014, 3, http://www.ncsl.org/Portals/1/Documents/forum/Forum_2014/Income_Inequality.pdf.

Chapter 23

1- "Number of monthly active Facebook users worldwide as of 1st quarter 2018 (in millions)," Statista – The Statistics Portal, accessed May 19, 2018, https://www.statista.com/statistics/264810/number-of-monthly-active-facebook-users-worldwide/.

2 – Ben Popper "Google announces over 2 billion monthly active devices on Android," *The Verge*, May 17, 2017, https://www.theverge.com/2017/5/17/15654454/android-reaches-2-billion-monthly-active-users.

3 – Brin, "2017 Founder's Letter."

4 – "Number of active Amazon customer accounts worldwide from 1st quarter 2013 to 1st quarter 2016 (in millions)," Statista – The Statistics Portal, accessed May 19, 2018, https://www.statista.com/statistics/476196/number-of-active-amazon-customer-accounts-quarter/.

5 – Casey Chan, "This Map Shows the Size of the World's Biggest Empires from History," *Gizmodo*, August 2, 2016, https://sploid.gizmodo.com/this-map-shows-the-size-of-the-worlds-biggest-empires-f-1784734014.

6 – Cindy Wooden, "Global Catholic population tops 1.28 billion; half are in 10 countries," *National Catholic Reporter*, April 8, 2017, https://www.ncronline.org/news/world/global-catholic-population-tops-128-billion-half-are-10-countries.

7 – "The Future of World Religions: Population Growth Projections, 2010-2050: Hindus," Pew Research Center, April 2, 2015, http://www.pewforum.org/2015/04/02/hindus/.

8 – Itamar Eichner, "Jewish worldwide population in 2015 is nearly 16 million," Ynetnews, June 26, 2015, https://www.ynetnews.com/articles/0,7340,L-4673018,00.html.

9 – Greg Kumparak, "Elon Musk Compares Artificial Intelligence To 'Summoning The Demon'," TechCrunch, October 26, 2014, https://techcrunch.com/2014/10/26/elon-musk-compares-building-artificial-intelligence-to-summoning-the-demon/.

10 – Media Ajir, "Effects of Transhumanism on United States National Security," (graduate studies, University of Nebraska-Omaha, July 2016), https://www.researchgate.net/publication/317084323_Effects_of_Transhumanism_on_United_States_National_Security.

11 – Russia Insight, "Whoever leads in AI will rule the world!- Putin to Russian children on Knowledge Day," video, 0:48, September 4, 2017, https://www.youtube.com/watch?v=2kggRND8c7Q.

12 – Greg Allen, "Thank Goodness Nukes are so Expensive and Complicated," *Wired*, March 4, 2017, https://www.wired.com/2017/03/thank-goodness-nukes-expensive-complicated/.

13 – Brin, "2017 Founder's Letter."

The Pledge

1 – Question Everything, "The Asch Experiment," video, 1:57, April 11, 2013, http://www.youtube.com/watch?v=qA-gbpt7Ts8.